T0220216

WHY STRING THEORY?

WHY STRING THEORY?

Joseph Conlon

CRC Press
Taylor & Francis Group
Boca Raton London New York

CRC Press is an imprint of the
Taylor & Francis Group, an **informa** business

CRC Press
Taylor & Francis Group
6000 Broken Sound Parkway NW, Suite 300
Boca Raton, FL 33487-2742

© 2016 by Joseph Conlon
CRC Press is an imprint of Taylor & Francis Group, an Informa business

No claim to original U.S. Government works

Printed on acid-free paper
Version Date: 20151020

International Standard Book Number-13: 978-1-4822-4247-8 (Paperback)

Visit the Taylor & Francis Web site at
http://www.taylorandfrancis.com

and the CRC Press Web site at
http://www.crcpress.com

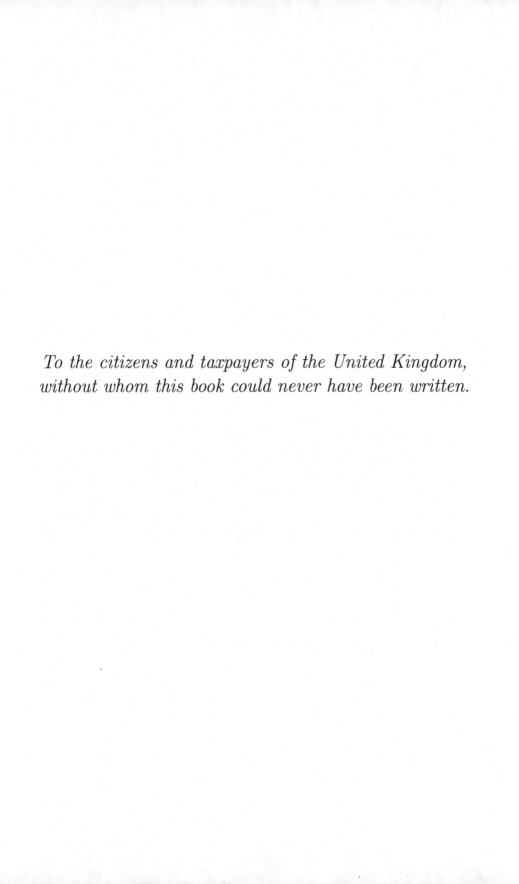

To the citizens and taxpayers of the United Kingdom,
without whom this book could never have been written.

Contents

Preface

This is a book about string theory. I started writing this book in 2010 with two motivations. The first was a desire to give back to those who have given to me – it is an incredible privilege to be paid to think about science and fundamental physics, and those who pay the piper have a right to hear the tune. The second was an intent to heed advice passed on to me since childhood: learn as much by writing as by reading. Writing is fun, and I wanted to go beyond the technical prose of journal articles.

This book is written in response to the questions of 'What is string theory?' and 'Why do you work on it then?', and it aims to answer them. String theory is, however, a large subject. Over its history, there have probably been over twenty thousand research papers on it. No one – and certainly not myself! – knows all of this work in detail. Writing a book requires an ordered account. In arranging material I have had to select topics, and in selecting I have aimed for the combination of intrinsic importance and personal familiarity.

To state the obvious, this is not a textbook. It does not have equations. It will not teach anyone either to do calculations in string theory or to perform research in the subject. There are many good textbooks that do precisely this, and I give references to some of them in the bibliography. What I hope the book does provide is a mental map of the subject, a tourist's guide to those who want to come and see the sights before returning to their own pursuits. I also hope that it may allow those who aspire to a career in the area a chance to reconnoitre the lay of the land.

To these ends, I have aimed to describe the physics as accurately as possible consistent with the constraints. Accuracy is not a single standard. The geography of the British Isles can be accurately described either as two triangles or through a bookshelf full of Ordnance Survey maps. Neither is 'right' or 'wrong'; it depends on the desired level of description. I have aimed my account of string theory somewhere between these two extremes.

The book is written to be read sequentially, but this is not an absolute requirement. A reader already familiar with the big picture of known physics could skip over the first part with little loss. In the third part, on the different applications of string theory, the chapters can be read in any order. The twelfth chapter, on the different styles of scientist, stands alone by itself.

One of the joys of physics is that it is a connected subject. Ideas repeat themselves, and on occasion in this book the same idea comes up more than once. When this happens I have sometimes allowed myself to re-explain concepts rather than expect the reader to remember an explanation from a hundred pages earlier. It always takes me multiple attempts to understand any concept, and in seminars I find few things more annoying than the speaker who believes that a single mention of an idea allows them to deny the audience any further reminders or clarifications of the topic.

When quotations occur within the book, I have gathered the details of the original sources into the end-notes and bibliography. Those who require them can find them, and I wish to avoid the academic custom of battering the reader into submission through a fusillade of references to obscure journals.

Many people have helped bring this book to fruition. I should first thank those, too numerous to mention, who have helped me professionally in learning the subject over the years. I particularly thank my fellow PhD students in Cambridge from 2003 to 2006, my research collaborators, and especially my doctoral supervisor Fernando Quevedo.

In the early stages of this book a website http://whystringtheory.com was made during summer 2012, constructed by myself and two undergraduate physicists, Edward Hughes and Charlotte Mason, who are now doing PhD degrees at respectively Queen Mary University of London and the University of California at Santa Barbara. I have borrowed the title of the book from the website.

Two eyes good, many eyes better: I thank those who have read parts of this book in draft form and whose feedback has helped me both sharpen the text and remove errors of physics, grammar and style: Lucy Broomfield, Frank Close, Theresa Conlon, Marcus du Sautoy, Pedro Ferreira, Sven Krippendorf, Fernando Quevedo, Markus Rummel, Andrei Starinets, David Tong, Peter West, and particularly Thomas Conlon, David Marsh and the anonymous copyeditor. The mistakes that remain are my own.

I also want to thank both my colleagues in the Oxford physics department and the Warden, Fellows, staff and students of New College, Oxford, for providing a stimulating and inspirational environment during the writing of this book.

I thank my editor at CRC Press, Francesca McGowan, for all her hard work, unfailing enthusiasm, and her many precise and detailed emails.

And finally, special and last thanks are reserved to Lucy, Alexander and George, for the joy they bring every day.

JOSEPH CONLON

New College, Oxford
September 2015

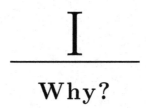

I

Why?

CHAPTER 1

The Long Wait

The 25th of June 1973 did not appear special. The Watergate scandal was rumbling on towards its denouement. In New York, the twin towers of the World Trade Center had opened two months previously. Nearby on Long Island, around the offices of the Physical Review, it was warm but not hot, with neither rain nor wind spoiling a pleasant summer day. The latest issue of *Physical Review Letters*, the prestige journal of the American Physical Society, had just gone to the printers. Among the articles in the issue that day were two on particle physics, appearing one next to the other. The articles addressed the same topic and had both arrived in the mail six or so weeks earlier. The first was by David Gross, a young 32-year-old associate professor at Princeton University, together with his first graduate student, 22-year-old Frank Wilczek. The second paper was by David Politzer, not much older at 23, who was then a graduate student at Harvard. The respective titles of the two papers were 'Ultraviolet Behavior of Non-Abelian Gauge Theories' and 'Reliable Perturbative Results for Strong Interactions'. Both papers addressed the same question: how the behaviour of certain physical theories varied when examined at large distances compared to when they were examined at small distances. The papers had been received at the journal's offices less than a week apart, and after favourable review were now appearing consecutively.

So far, so ordinary. However, from the perspective of forty-two years on, what is special about that summer's day in 1973 is that, at the time of writing, it represents the last time work in theoretical particle physics was published that would subsequently go on to be awarded the Nobel Prize for Physics.

There are four known forces in nature: the gravitational force, the electromagnetic force, the weak force and the strong force. The gravitational force is familiar, as are the effects of electromagnetism. The strong force serves to hold atomic nuclei together, but it is entirely absent in the macroscopic world. It remained mysterious for years how the physics of the strong force could be so important in the nucleus and so irrelevant everywhere else. This physics was even more mysterious during the 1950s and 1960s as a vast zoo of particles all interacting under the strong force were discovered, one after another after another. The behaviour of the strong force appeared bizarrely different to any of the others, and at the time it was thought the strong force might defy understanding for fifty or a hundred years.

It did not. Despite many opinions to the contrary, the equations underlying the strong force were in fact similar to those underlying the electromagnetic and weak forces, and the apparently inexplicable behaviour of the strong force was, surprisingly, understandable in terms of known theories. The work of Gross, Politzer and Wilczek was the key to showing this, and for this discovery they would, in the fullness of time, be awarded the 2004 physics Nobel Prize.

Since then, there have been more recent Nobel Prizes in particle theory, such as the one won by Peter Higgs and François Englert in 2013, but these have been for work done even earlier than 1973 – in 1964 for the case of Higgs and Englert. This has been neither for want of papers nor for want of effort. It is a stubborn truth that the theory that had just about been written down in the 1970s as an approximate description of nature – the Standard Model – has turned out to be far better than anyone could conceive or imagine at the time. The Standard Model is the modern description of all known elementary particles and their interactions, and it has survived all experimental attempts to prove it wrong. Tests that in the 1970s were either only qualitative in form, or even only an experimental dream, have morphed into successful high-precision probes of the fine structure of the Standard Model.

Over this period many proposals have been made for new particles and new phenomena beyond those already in the Standard Model. Indeed, there have even been some experimental claims to have found such particles. However, these claims of discoveries have not held up, and the proposals have failed empirical test. No such new particles have yet been observed. This history is full of many experimental triumphs and first discoveries of new particles – most recently the 2012 discovery of the Higgs boson at CERN's Large Hadron Collider – but these all involve discoveries of previously unobserved particles of the Standard Model. It is not impossible that some of these proposals will eventually turn out in some form to be correct at energies that are beyond current technology. There is however so far no positive evidence for this, and as new experiments replace old ones, the young bloods who with dreams of Stockholm wrote down ideas for new physics are now admiring the first hesitant steps of their grandchildren.

The big picture of the last forty years, then, is the consolidation of the Standard Model from a house of straw into an enduring edifice of high-energy physics. Over this long period, this fact has led to two broad directions of thought. The first has been towards bettering our understanding of the Standard Model. This has involved answering in ever greater detail the questions: what are the predictions of the Standard Model? What is the most efficient way to work out these predictions? At first, no one could have realised that this was the path they were headed down. The Standard Model originally appeared a temporary ersatz construction that would be good for a few years until something better came along. Understanding the predictions of the

Standard Model and discovering the new physics that went beyond it did not appear to be separate enterprises – they were different sides of the same coin. However, with the passage of time the machinery required to compute the predictions of the Standard Model has grown increasingly elaborate, and the tests that are used to look for new physics require ever greater precision. To say something new about the Standard Model, it is no longer sufficient to be an expert – one must now be an expert even within the community of experts.

The other direction has been towards deep exploration in the empty quarter of theory space. The direct empirical motivation for many of these peregrinations is somewhere between marginal and non-existent. There is little or no attempt to use these wanderings either to explain current experimental anomalies or to make predictions for experiments on a near-future timescale. It involves theoretical structures that are inspired by those that are observationally relevant, but the inspiration may be loose. The aim is to understand these theoretical, perhaps quasi-mathematical, structures and to grasp their inner mechanics. Such theories may either never be applicable to nature or may only apply at energy scales reaching far beyond those accessible in current particle colliders. The close connection between advances in theory and advances in experiment is put aside in favour of the deep study of theories for their own sake, with mathematics rather than experiment as the closest companion in the attempt to advance the subject.

One of the best-known finds from these journeys is the subject of this book, string theory. Over a period of almost fifty years from its origins in 1968, string theory has spread from nothing to become a major component of theoretical particle physics. In terms of money, theoretical physics is not a big subject. It is a flea-bite on experimental particle physics, which is itself a flea-bite compared to medical research, which is in turn a flea-bite on the totality of government spending. Yet intellectually, particle theory has always punched far above its financial weight. Its topic is the basic laws governing our universe, and interest in these laws extends far beyond the confines of those who are salaried to think about them. Theoretical physics has provided – certainly not all, but definitely one part – of the answer to the universal questions of who we are and where we come from.

It is also simply true that within particle theory, string theory is a big subject. There are few large particle theory groups within major universities that do not have at least one person doing string theory. In total, there are probably a couple of thousand people at universities around the world whose mortgage is paid by either doing string theory, using string theory, working with tools made within string theory, solving problems using methods developed in string theory, or simply having their mental map of the world at the smallest possible scales set by string theory.

String theory is most famous as a theory of quantum gravity and a candidate theory of fundamental interactions at the smallest possible scales, scales that are as small compared to an atomic nucleus as an atomic nucleus is to a person. These scales are also far smaller than can be directly probed by any experiment, even the Large Hadron Collider at CERN. This raises a natural question – if you cannot test string theory directly, how do you know it is right? And if you do not know it is right, and you will not know anytime soon, why do so many people work on it? Surely science advances by experiment, and without experiment there can be nothing but stagnation.

This is a question posed by many decent, smart and inquisitive people. They may be undergraduate students of physics; they may be evolutionary biologists; they may be observational astronomers; they may be intellectually curious members of the general public who are excited about the fundamental laws of physics. Those who ask this question are people of good will and good faith, who in all charity cannot understand how any one scientific idea without direct experimental support could ever command the time and allegiance of such a large fraction of the relevant experts.

For it is true that, the closer you approach the subject professionally, the fewer and fewer people you find who continue to ask these questions. Scepticism about string theory tends to diminish as technical knowledge about the subject increases. In practice, the hard acid that dissolves scepticism is the ability to calculate. The greater the ability to calculate in theoretical physics, the clearer the appeal of string theory becomes, both for its own intrinsic perfectly fitting calculational structure and also for its ability to shed light on results in other areas. However it is also true that in return for this ability to calculate, one must pay dearly in terms of time, effort and earning ability. Most people, however strong their interest or aptitude, are for perfectly good reasons not able to follow this route.

This book is written for these people and aims to answer the questions above. I want to explain, to my wonderful fellow citizens who support scientific research through their taxes, why string theory is so popular and why, despite the lack of direct empirical support, it has attained the level of prominence it has. I want to explain how string theory fits into the broader picture of science, and just why experts find it so compelling. If all the varied rationales of this book were condensed into one single aim, it is to answer the question: why do so many experts choose to work on a theory that has no direct experimental support? My dear colleagues know the answer already, but this book is not written for them.

This is also, for obvious reasons, a book (almost) without equations. There exists an opinion that if you have true understanding, equations are unnecessary and you can convey concepts with no losses in transmission. I do not share this view. Mathematics is the light that illuminates physics, and mathematics-free physics has marked similarities to alcohol-free wine. Mathematics is the most powerful tool available for explaining and understanding physics, being to a physicist what fingers are to a pianist. Nature really is written in the language of mathematics, and many results that flow easily using it are hard to justify in any other way. In this subject mathematical training is a pure and unqualified good: there is no additional raw creative freshness available through its absence.

So I apologise now for the fact that the argument made in this book is only a shadow of the argument I would like to make. I am also sorry, despite my best efforts, for the parts of this book that appear less than clear, or where the logical jumps seem broad and unjustified. That said, that which is worth doing is also worth doing poorly. One does not need the palate of Carême to enjoy food, or the ear of Barenboim to listen to music. The rose windows of Chartres are both for those who can follow the biblical stories there and for those who cannot – and the great structure of physics, at least in its outline, is part of the inheritance of all interested minds.

I now want to describe the structure of the book and the overall organisation of its argument. The first part of the book aims to place string theory – the theory of quantum mechanical, relativistic strings – within the broader context of both science as a whole and more specifically theoretical particle physics. There is a lot of science – most of science, indeed almost all of science – that has nothing to do with either the fundamental laws of nature or the discovery of new physical principles. My first aim is to isolate the small part of science for which string theory is potentially relevant, and then to describe the important truths that are already known in this area. These topics are covered in chapters 2 and 3. Chapter 2 deals with science as a whole, classifying it by the different length scales that are appropriate. Chapter 3 focuses specifically on physics and on the confirmed results that have been used to build up the modern picture of both particle physics and cosmology.

Whatever string theory may be, it is certainly outside the canon of established and experimentally confirmed theories. Why do we even need anything new at all? Chapter 4 makes the case for this. It argues that what we have is not enough: despite all previous successes, there is still a need for ideas and structures that go beyond what we already have. When we look at what we have learnt, we see that indeed it is very good – but it does not suffice. Chapter 4 makes no argument for string theory as such. It instead argues for something new. It argues that there must exist something that is not included in our existing theories but instead sits above and beyond them. It explains why theory, observation and experiment all demand that what we currently know is not ultimately at the top of the intellectual food chain – we are still missing a Godzilla theory that is the top predator.

String theory is, in the first instance, a candidate for that something. It is introduced in the second part of the book and is the subjects of chapter 5 and 6. These are devoted to explaining first what string theory was, and then what string theory is. What do the words 'string theory' actually refer to? What does the expression 'string theory' even *mean*? This is a simple question with a complicated answer that has changed dramatically over the last fifty years. What 'string theory' meant in 1970 was not what it meant in 1985, which in turn was not what it meant in 2000. Likewise, the motivations for working on the subject have also changed over these periods. The reason for first studying string theory in 1970 was a very bad reason for studying string theory in 1985 but had returned to be one of the most common reasons in 2010. While the underlying equations and calculations have undergone a mostly smooth development, the conceptual picture of what string theory is and how it fits into physics has changed dramatically.

These chapters are therefore both explanatory and historical. To appreciate what 'string theory' means in 2015, it is helpful to know what 'string theory' had previously meant in both 1970 and 1985. Over this period, the history of string theory interleaves with the history of particle physics. Chapter 5 is concerned with past understandings of string theory, while chapter 6 presents a more current view.

Having reviewed the history and nature of string theory, the third part of the book describes motivations for doing research on it. Chapter 7 is devoted to the external correctness of string theory. It reviews the direct experimental evidence proving that string theory is the correct theory of nature at the smallest possible scales.

The following four chapters explore several particular topics in further depth. The essence of these chapters is that scientists work on string theory because string theory is useful to them. The human foibles of scientists are often appreciated more in the abstract than in the reality. Scientists have their own pet topics, and they care most about what will assist them in their own passions. One of the dirty great secrets about 'string theory, the candidate fundamental theory of everything' is that the majority of people who work on this subject actually care little for strings *per se* and absolutely nothing for fundamental theories of everything. Instead, they have their own interests, which are generally not formulated in terms of string theory, and they care about string theory because it can help them understand what they are actually interested in. The third part of the book describes this in detail and explains why people really choose to work on string theory.

Chapter 8 is on quantum field theory. We do not know whether string theory is the correct theory of nature. We do however know that the Standard Model is a correct theory of nature, and the Standard Model is built up by describing the strong, weak and electromagnetic forces through what is called a quantum field theory. Quantum field theories are a known part of nature, and they are used not just in particle physics but also in the physics of matter, to describe the way quantum behaviour manifests itself in macroscopic bodies such as insulators or conductors both semi- and super-. This ubiquity of quantum field theory causes many physicists to be interested in understanding it better. They care both about special examples where a quantum field theory can be solved exactly, and also about general insights into how to understand and work with quantum field theories in regimes where all normal techniques break down. Indeed, probably the most popular reason to work on string theory in the years since the millennium has been for the help it provides in understanding quantum field theory.

Another reason for working on string theory, covered in chapter 9, stems from an interest in mathematics. The developmental logic of string theory was that of physics: it has been constructed by physicists using the techniques and standards that apply in theoretical physics. Nonetheless, it involves geometric ideas and the subject has many overlaps with topics of interest to mathematicians. This creates a fortunate situation – it offers a perspective that is removed enough to be novel but close enough to be useful. The logic of physics is different to that of mathematics, and what is obvious in physics is not obvious in mathematics, enabling the discovery of results that are both new and striking. Since the middle of the 1980s, string theory has acted as tinder for the historic romance of physics with mathematics, provoking both new mathematical techniques and new approaches to existing problems.

Of course, many people who work on string theory (including me!) do so because they are fascinated by particle physics and cosmology and want to know what lies beyond our existing boundaries of knowledge. This is the topic of chapter 10. String theory provides a rich structure that links onto what we already know about physics while also offering many suggestions for what may lie beyond. While it is implausible to claim that string theory offers any unique extension of the Standard Models of particle physics and cosmology, it provides plentiful examples of types of new physics that one would not otherwise think about. In this respect, string theory is a fine muse and a fertile source of ideas for new ways of thinking about new physics. String theory may not easily be definitively tested, but it is a good source of ideas that can be. These ideas then enter the conventional process by which experiments can be proposed and built to check their veracity.

Chapter 11 deals with string theory as a theory of quantum gravity. This is the headline application of the subject and has indeed seen a significant amount of work over the years. The chapter describes both the reasons for believing that string theory offers a quantum description of the gravitational force, as well as specific applications to some of the classic problems of classical and quantum gravity.

The fourth and final part of the book discusses the social aspects of science. The public view of string theory and string theorists has fluctuated over time. One perspective has seen string theorists as the intellectual equivalents of John Wayne, using the power of pure thought to conquer the untamed territories of physics. A more recent outlook sees them as surrender monkeys who inherited the great tradition of theoretical physics and fled from experiment for a combination of mathematical arcana and personal wonga. I hope to show that the truth is more human than the first and richer and more powerful than the second.

One of the overarching themes of the book is that of diversity. Different scientists care about very different topics, and there are as many approaches to physics as there are physicists. Geoffrey Boycott and Kevin Pietersen were both excellent cricketers who garnered a sackful of runs, despite having personalities and playing styles that could not have been more different. The same holds true in physics. Some prefer deep and intricate calculations with a clear right answer that emerges only after many pages of precise and exact work. Others prefer drawing big picture connections between disparate areas with no apparent relation. Some are only happy when working in close tandem with experiment, while others find satisfaction in the pristine purity of mathematics and are repulsed by the dirtiness of data. There are many ways to be a first-rate physicist and no one choice is best. Chapter 12 is devoted to pen portraits of some of these different styles of doing physics.

Chapter 13 considers criticisms of string theory. String theory is discussed in fora that reach far beyond the circles of theoretical physics, and it is a tribute to the culture and education of our society that many people care about aspects of fundamental physics so removed from their daily lives. In the same way that honest and engaged citizens can have markedly opposed political views, attitudes to string theory vary sharply, and some scientists are passionately concerned that string theory is damaging physics. This chapter aims to take these attacks, frame them in their most convincing form, and reply to them.

The final chapter summarises the overall message of this book: why is string theory so appealing? There are other speculative theories of physics at the smallest possible scales. This chapter explains why string theory has been so much more successful than them. What makes string theory special? The short answer is that, in contrast to other approaches, string theory is so much more than just a candidate theory of quantum gravity.

Other theories of quantum gravity are precisely that – other theories of quantum gravity. In contrast, string theory is a theory for the pragmatic as well as for the ideological. It has outgrown its origins and has something to offer those who are completely uninterested in quantum gravity. Its success in solving problems and contributing deep insights to so many topics has made it popular with many; the same deep insights, far away from its native subject, also convince many that it is the most plausible idea on its home ground. String theory has given much to many, and its enduring popularity comes from the fact that, over several decades, it has remained a theory that just keeps on giving.

CHAPTER 2

Scales of Science: Little and Large

'As-salaam alaikum, from where are you coming and to where do you go?' The ancient call of the desert, from caravan to caravan as they passed on the spice road, also applies to science. A subject with no past can lack the techniques necessary to do science. A subject with no idea as to the problems it is trying to solve has no future.

This is a book about string theory. Where does string theory fit into the larger canvas of science and physics? 'Science' is a big tent with many tent dwellers. Its discoveries have been the primary facilitator of the enormous economic growth of the last few centuries, and few politicians fail to pay at least lip service to it. In the United Kingdom, the annual government budget for science is around ten billion pounds,[1] although the overall sum spent is slightly larger because of charitable funding of medical research and the industrial research done in large pharmaceutical and engineering companies. In this big friendly tent, where does string theory fit in? The short answer is: not where the money is. In the most recent academic year, string theory research received somewhere south of one tenth of one per cent of total science funding in the United Kingdom. In the affairs of pounds, shillings and pence, string theory is a tiny tiddler even within a relatively small pond.

Cost represents one way of classifying different branches of science. Another more subjective measure is by practical utility. String theory belongs clearly to those branches of science that are the canaries in the coal mine for curiosity-driven research. These branches are pursued for the value of intrinsic understanding rather than for any vision of commercial application. As with astronomy or particle physics, string theory is clearly pure, rather than applied, science.

---

[1] As there are different ways to measure it, this number should not be taken as overly precise.

It is true that the pure science of one generation has tended to become the applied science of the next. The quantum theory of the electron underpins the semiconductors and transistors that sit behind all modern electronics. The accuracy of a GPS device relies on Einstein's laws of general relativity. 'Proton therapy', the poster child of advanced cancer treatments, is applied particle physics. However, whatever the unknown future may be, it is important to be clear upfront that there are no currently foreseeable economic applications for string theory.

We will arrange our scientific taxonomy differently, classifying sciences according to their characteristic distance scales – the size of the phenomena they study. 'Characteristic' is a wooly word that is nonetheless more useful than an overly exact definition. Its usage is best illustrated by example. The characteristic size of a blade of grass is a couple of centimetres. The characteristic size of a mountain is a couple of kilometres. The characteristic size of a galaxy is a hundred thousand light years.

We are all human. Excepting hobbits and point guards, an adult member of *homo sapiens* is typically between 150 and 190 centimetres tall. He or she is made predominantly of water and weighs between forty-five and ninety kilograms. These basic facts determine our first scientific, or rather proto-scientific, intuitions. Our growing lives are conditioned by and organised around the effects of gravity. Children learn at an early age that they cannot walk up walls and that if they fall over they hurt themselves. These 'natural' effects are natural only due to our characteristic size. Intelligent ants would experience the world very differently. Gravity is far harder to discover if you can walk up a wall without difficulty and fall down a mineshaft without danger. There is not that much ant for gravity to act on, and it is surface tension, rather than gravity, that plays the crucial role in daily myrmecological life.

We are all human. We measure distances in metres and time in seconds. It is not a coincidence that the acceleration due to gravity, measured in these units, has a numerical value of ten metres per second per second – not a thousand, and not a thousandth. Metres and seconds are units that are adapted to us, which is why our most familiar force is characterised by a number not too different from one when measured in these units.

From our own scales, there are two ways to go: up and up, or down and down. As we go up, we increase the distance scale. The metre becomes the kilometre and we step beyond the scales of living organisms. These scales belong to geography and geology, and they are associated with a profound if prehistoric discovery: the earth is not flat but round.[2] The evidence for this is simple: as ships sail to distant ports, they disappear beyond sight over

---

[2]There exists a lazy but widespread prejudice that this discovery was not prehistoric. The beliefs that the geography of the earth was thought not to be spherical and that Christopher Columbus was warned not to sail because he would fall off the edge of the earth: these myths descend from the Victorian fallacy that there was a mediaeval fallacy that the world was flat.

the horizon. First the body of the ship disappears, and finally the masts. This effect is due to the curvature of the earth. Attempting to see a distant ship is like attempting to see over a hill. From the distance at which a ship disappears, it is possible to estimate the radius of the earth. Indeed, already in the third century BC the Greek scholar Eratosthenes – using a different method – made an accurate measurement of the diameter of the earth. The sphericity and size of the earth is the first major result in astronomy. Islands, countries and seas have distance scales ranging from kilometres to thousands of kilometres, and it is these scales that are used to characterise the geography of earth. For example, the distance to the centre of the earth is around six thousand kilometres, and the tectonic plates on which the surface of the earth is supported are thousands of kilometres across.

Larger distances are not elegantly described in terms of kilometres. The distance from the earth to the moon is around three hundred and eighty thousand kilometres. The distance from the earth to the sun is around one hundred and fifty million kilometres, while the distance to the next star is around forty thousand billion kilometres. The numbers become clunky, and do nothing to develop intuition. There is a more natural kind of unit: light units. These involve the distances travelled by light in either seconds, minutes or years. Light travels at approximately three hundred thousand kilometres a second. One light second corresponds to a distance of three hundred thousand kilometres while one light year corresponds to a little over four thousand billion kilometres. Within the solar system, the sun is a mere eight light minutes from earth, and the earth just over one light second from the moon. The planetary denizens of the solar system – Uranus, Mars, Venus and Jupiter – took their names from the Roman deities; now they are simply the domestic avatars of planetary science.

The next step is from one star to many: as light minutes become light years, we move from our own sun to all other stars, and the scale of the galaxy is reached. Galaxies are a collection of stars, some very much like the sun and others less so. This collection numbers around a hundred billion stars in total. These all circle an enormously heavy black hole at the centre of the galaxy, in the same way that the planets circle our sun. The size of our galaxy is such that it takes light around one hundred thousand years to cross it, and such that a galactic year, the time taken for the sun to rotate around the galaxy back to its current location, is two hundred and twenty-five million years. By now these distance scales, the scales of galactic astronomy, are so large that we have no real human conception of them. We can state them, we can calculate with them, but we cannot feel them.

For one science even this distance is parochial. That science is cosmology, the study of the whole universe. A cosmologist asked to discuss galactic dynamics is like a national politician asked about the church fête: the event is rather parochial for those with eyes on grander matters. Cosmologists concern themselves not with thousands of light years but billions, and not with single galaxies but the statistical properties of thousands: how far apart are galaxies on average, and how do they cluster together?

Coming up from human scales, the scales attained in cosmology represent the largest distances in science. There are no bigger scales known, and we literally cannot see any further. There is a finite amount of time since the Big Bang, and in this period light has only had the opportunity to travel a certain finite, albeit large, distance. Light is what we can see, and light travels further and faster than anything else.

This ends the upwards journey. Before asking where string theory fits in, we also want to trace distances downwards, from the scales of humans down to the smallest known scales. It turns out, surprisingly, that the science of the very largest distances is closely related to the science of the very smallest. As for the far left and far right in politics, the principles that apply at the extremes turn out to be remarkably similar to one another.

We started the upwards journey on the human scale, the scale of metres, and we return here for the downwards journey. This scale is shared by other large mammals and is the largest of the biological scales – even the biggest of the dinosaurs was never more than tens of metres in length. The reason for this is gravity: the bigger you are, the more energy is required to move muscles against gravity. The largest mammal today is the blue whale, which only achieves its size by mitigating these effects through the buoyancy of a watery home.

The large animals are merely the most biologically visible, sitting at the top of the digestive ladder of who eats whom. As we come down in distances by a factor of a hundred, we start measuring lengths in centimetres and millimetres. This is still the scale of biology, but it is the scale of a different kind of biology: ants and other insects live here. Why is this a different kind of biology? This is because, as mentioned above, clever ants and clever insects would develop science in a very different order to us. They experience the world differently, and the differences are entirely due to size. Where horses can only splash, insects such as pond skaters can actually walk on water. This is so because they are small enough that the surface tension of water can dominate the effects of gravity. Unlike elephants, small insects can be picked up and blown away in the wind, and they can fall great distances without harm. An ant weighs one hundredth of a gram and a human one hundred kilograms. The gravitational force on an ant is almost a billion times weaker than that on a human, and gravity plays a proportionately smaller part in the entomological environment.

Ants and other insects do, however, have eyes, which they use to navigate their world. The next step on our descent is to the largest end of the so-called mesoscopic scale, involving distances comparable to one micrometre, one thousand times smaller than a millimetre. One micrometre is approximately twice the size of the wavelength of visible light. Distances smaller than this cannot, even in principle, be resolved using ordinary light; it would be like attempting to make filigree lace whilst wearing boxing gloves. This scale is the end boundary of biology, inhabited by the smallest living organisms. A single cell is around ten micrometres across, and a single virus around one-fifth of a micrometre in length.

The mesoscopic scale extends down to tens of nanometres. The mesoscale is also the scale of the physics of polymers and liquid crystals. The physics of these distances is one where gravity is entirely negligible and all forces and interactions arise from electromagnetic effects and the distribution of electric charge. It is at this scale that the fictional forces, such as friction, are seen to originate from interactions between the surfaces of two bodies. I use the expression 'fictional' not because friction does not exist, but because it does not have an independent existence. 'Friction' is a macroscopic name given to the effects of microscopic interactions. These electromagnetic interactions are remarkably strong. One of the easiest ways to understand this is to note that in a sitting position the downwards gravitational pull of the entire earth – all thousand billion billions tons of it – on the bottom is defeated by the short range electromagnetic effects between the chair and the derrière. While these short-range forces are known by many names – 'stress', 'strain', 'normal reaction' and 'friction' – each one of these only serves to capture a different macroscopic aspect of the electromagnetic forces of the mesoscale.

As we pass through the mesoscale, we cross the great Rubicon of physics. We start with distances described using classical physics, for which there is no need to introduce quantum mechanics. We end with distances for which there is no longer a good classical description – the world is quantum, and irrevocably quantum. This marks a radical change in how we can think about physics, as an object can no longer have both a well-defined location and a well-defined speed. Quantum mechanics is subsequently needed at all distance scales smaller than this, and it is never again absent from the equations.

The next big step down is to the nanoscale at one-billionth of a metre. Once at the nanoscale, the truth that the world is quantum and not classical is undeniable and encountered everywhere.

The nanoscale is also the scale of atoms. On one level, atoms are tiny. They were traditionally viewed as the fundamental building blocks of matter, and their name comes from the Greek for indivisible – 'a-tomos'. Atoms are also at a scale so far removed from everyday life that the existence of atoms was not established until the 19th century. Size is however relative, and the atomic scale is simply a staging post on our continuing journey. It is easy to think that as atoms are small, everything within them is equally small. This is not so. Atoms consist of electrons with negative electric charge orbiting a tiny central nucleus, which has positive electric charge and consists of protons and neutrons.

One of the greatest of all conceptual realignments in science was the discovery that the periodic table of elements, and their properties, follows entirely from quantum mechanics applied to protons, neutrons and electrons. The hydrogen atom is the quantum mechanical solution of the system of an electron and a proton. The helium atom is the quantum mechanical solution of the system of two electrons around a nucleus consisting of two protons and two neutrons. The scale of atoms is the scale of chemistry, the science that studies the combination of atoms to form molecules, and the subsequent interactions between atoms, molecules and other molecules. Chemistry is 'just' applied quantum mechanics: the 'chemical bond' is a name that describes the effects of the solutions of quantum mechanics.

In the hydrogen atom, the proton is a minute central ball of charge while the electron is diffused throughout the atom. Quantum-mechanically, the electron is everywhere. The nucleus, even quantum mechanically, is confined in a tiny central region. Tiny here really does mean tiny. To get a sense of the relative scales, if the hydrogen atom as a whole were blown up to the size of a mountain, the central nucleus would still only be the size of a lightbulb.

The laws of quantum mechanics govern the diffuse, probabilistic motion of the electron around this central lightbulb. What happens within it? The laws of quantum mechanics also govern the far more complicated internal behaviour of the lightbulb. The step from the scale of the atom to the scale of the nucleus brings us to the scales of nuclear physics. Nuclear physics is a science with a glorious infancy in the 1930s and 1940s, a happy maturity in the fifties and sixties, and a long gradual decline since then. The atomic nuclei are made of some number of protons and some number of neutrons. There are many different nuclei, which can be enumerated as a botanist enumerates flowers. These different types of nuclei directly give the diversity of chemical elements. The distinct chemical elements directly correspond to the different numbers of protons that a nucleus can have. The interactions *between* atoms are determined by physics at the scale of the atom; the distinct *types* of atom are determined by physics at the scale of the nucleus.

Nuclear physics blossomed as a subject in the 1930s, when it was understood that nuclei of one type could turn into nuclei of another type. This could occur through occasional random processes – radioactive decay of either alpha or beta kind, as first identified by Marie and Pierre Curie. It could also occur through human intervention. It thus became apparent that the base metals were not immutable, and the Philosopher's Stone that had been sought for millennia was realised as a focussed beam of particles powered by a Cockroft-Walton accelerator. It was also clear that a deeper layer of physics underlay the changes between the different nuclei.

This deeper layer of physics remained mysterious for a long time. Nuclei, made of protons and neutrons, were known not to be fundamental objects. Complex nuclei consist of large numbers of both protons and neutrons – a typical nucleus of gold contains seventy-nine protons and one hundred and eighteen neutrons. Atoms and their nuclei therefore joined earth, air, fire and water as non-elemental objects that, on close examination, fall to bits.

However, even single protons or neutrons are composite, rather than elementary, objects. A free neutron, left to itself, will decay on an average timescale of fifteen minutes, leaving a proton behind. If a proton is hit hard enough with another particle, it will also break apart. Protons and neutrons, each with a radius of around one femtometre – one millionth of a nanometre – are the smallest known composite objects in the universe.

The science that deals with distances smaller than this is particle physics. This science describes elementary particles such as the electron – and also the quarks that make up the proton and neutron. There are three quarks required for both a proton and a neutron, held together by the strong force. The name quark is an odd one and comes from a literary indulgence: originally called 'aces', the human encyclopedia and Nobel Laureate Murray Gell-Mann gave them the name that stuck after a word from James Joyce's *Finnegan's Wake*. As far as is known, the quarks join the electron as elementary particles with no further substructure.

Particle physics is, as it says on the tin, about the physics of particles. At its most basic level, it involves enumerating the known particles, searching for new ones and determining the laws and rules that govern their interactions. 'Enumeration and classification' is by itself no more than taxonomy. The real thrust of the subject, capable of waking its practitioners in the night, lies in the search for the deep truths that organise hundreds and thousands of different observations and measurements into different facets of a single beautiful idea. Particle physics may appear complex, but at heart it is a simple subject for simple minds. It is a simple subject, but not an easy one. Its simplicity is of the kind that lies on the far side of difficulty, confusion and complexity. The journey is difficult, but the view once achieved is magnificent.

The known range of particle physics extends down to around one ten thousandth of the size of the nucleus. It has achieved this by first accelerating particles to high energies and then colliding them. The larger the energy, the smaller the distance scales probed. The most energetic collider in the world is the Large Hadron Collider at CERN outside Geneva, which sets the current experimental frontier.

In a completion of the upwards and downwards journeys, it turns out that physics at these very smallest scales is also crucial for cosmology, and in particular for the distribution of matter throughout the visible universe. In a sense that is not totally inaccurate, the Large Hadron Collider can be said to recreate the conditions that held in the very earliest epoch of the universe.

Over the last few pages, I have described the sciences of many different scales. These are all scales on which experiments can be carried out, and almost all of 'science' is contained within one of these scales.

String theory is not the science of any of the above distance scales. It is directly relevant for the more hypothetical regions where one asks: what happens as distances get smaller and smaller? This is a real part of nature, but not a part where it is currently possible to do experiments. It involves extrapolating from what we know to be true to what might be true.

What are the laws of nature that apply at these shorter distance scales? At the smallest scales we can measure, particle physics tells us that the laws of nature are described by the combination of quantum mechanics and special relativity. The only consistent way of combining these is through what is called quantum field theory. We first ask: can these techniques apply down to arbitrarily small distances? We know that all previous laws of physics have either broken down or required modification as they have been pushed to smaller distances. Nonetheless, if the experimentally confirmed ideas of particle physics would all also work there, Occam's razor – the statement that entities should not be multiplied without necessity – could be taken to suggest that there is no pressing need to investigate radically new ideas.[3]

One motivation for string theory, and one reason why this book exists, is that we know that this is not true. What has worked so far must fail at sufficiently short distances. In fact, the theories that we currently have signal their own demise. As we move continually down in size, there is a distance by which this failure must become manifest – a gaping and obvious hole in our attempts to understand nature. This distance is called the Planck length. This distance is genuinely tiny: one billion billion times smaller than the typical radius of the nucleus. It is as small compared to a nucleus as an atom is compared to the earth. This distance is nonetheless finite, and is a distance scale by which we know that our knowledge of physics must break down.

Why must current physics fail? We shall return to this in more detail in the next two chapters. Here, let me say that particle physics describes quantum particles in a classical spacetime. Spacetime is the fixed classical background against which the quantum interactions of particles take place. However, Einstein taught us that spacetime is actually not fixed – spacetime interacts, both with particles and with itself. The fully sophisticated version of this is general relativity, for which Newtonian gravity is for almost all purposes a good approximation.

The approximation of fixed spacetime is equivalent to neglecting the effects of gravity. For most purposes in particle physics this works wonderfully, because gravity as a force is so, so weak: using only my hands I can pull myself up and defeat the gravitational pull of *the entire earth*. The Planck length is defined as the distance scale at which this approximation must finally fail. At these scales, the effects of gravity can no longer be neglected, and the quantum theories of the electromagnetic, strong and weak forces must be joined by a quantum theory of the gravitational force.

There is also an important point to make about the Planck length. The Planck length tells us the smallest possible length that the theories we know and love *could* be correct down to. It does not tell us the length at which these theories actually do break down. It is entirely possible that these theories fail on lengths considerably larger than the Planck length. While everyone agrees that what we know must fail by the time we have reached the Planck length, no one knows exactly where, between the scales we have probed experimentally and the Planck length, the great engine of known physics judders to a halt.

---

[3] William of Occam, a mediaeval Franciscan friar, studied in Oxford around 1300, possibly within some of the same buildings that are still used by the university today.

A theory of the Planck scale requires a correct quantum mechanical description of gravity. Obtaining a definitively correct quantum theory of gravity is possibly the biggest open problem in theoretical physics. Cometh the hour, cometh the solution. And cometh another, and another, and another. Just as national crises see more than one politician volunteering themselves as the saviour of the country, so the important and well-posed nature of the quantum gravity problem sees many ideas put forward as proposed solutions.

However, something new must come in at the Planck scale. What will that something be? Modern theoretical physics says – if not by unanimous agreement then at least by a large majority – that something is most likely to be string theory. String theory states that somewhere between the scales of the Large Hadron Collider and the Planck scale new degrees of freedom will come in, and those degrees of freedom will be the higher harmonics of vibrating strings.

It should be said here that consensus opinions are dangerous things. There is a consensus opinion that the world is not flat, but there also used to be consensus opinions that smoking did not cause health problems and that there was no need to sterilise medical equipment between separate operations. Consensus opinions about the Planck scale deserve a heightened level of scrutiny because of the great difficulties in evaluating any such ideas experimentally.

The simplest way to describe string theory, then, is as one candidate solution of the quantum gravity problem, and so as a candidate theory of physics at the smallest possible length scales. As the book will show, this is a rather limiting definition of the subject. If all string theory had to offer was a possible idea for quantum gravity, it would receive far less attention than it does. However, that argument is one I intend to develop over the rest of the book, at the same time showing that the consensus opinion about the importance of string theory is held for good reasons.

The purpose of this chapter has been to locate string theory within the scientific vista. In this picture, its natural home is among the physics of the smallest possible scales. The ultimate story is richer, and we will later see surprising connections to physics at much larger distance scales, but this is where string theory has arisen from. The techniques it uses are most closely related to those of particle physics, with which it shares much of its language and personnel. Before narrowing down onto string theory however, I want to first review some of the great truths of physics.

CHAPTER 3

Big Lessons of Physics

3.1 SPACE AND TIME ARE CUT FROM THE SAME CLOTH

'In the beginning . . . ' History, culture and religion – as in the first words of the Bible – reflect a classical notion of time that is easy to understand and appears obviously correct. This notion says that there exists one, universal time that applies everywhere and to everyone. It has counted from the beginning to now, and it counts from now to the future. The lives and fortunes of mice and men are marked out by the beat of a celestial gong. This view was expressed by Isaac Newton in the *Principia*, the work describing his laws of motion:

> 'Absolute, true, and mathematical time, of itself, and from its own nature flows equably without relation to anything external, and by another name is called duration; relative, apparent, and common time is some sensible and external (whether accurate or unequable) measure of duration by the means of motion, which is commonly used instead of true time; such as an hour, a day, a month, a year.'

According to this view, we could in principle synchronise our watches with distant aliens visiting from a far-off galaxy. The aliens could return to their home galaxy, stay there for many years, and then once again come to Earth. On their arrival we could compare our watches once more, and we would find them to be still identical. In this picture, time sits above and apart from space, acting as a universal coordinate labelling events in space. This notion of time is so ingrained in us that it took a very special person to realise that it is in fact wrong.

That person was Albert Einstein, and the realisation occurred in 1905 with the development of special relativity. Special relativity is sometimes crudely described as the addition of a fourth dimension, with that dimension being time. If this were all special relativity was, Einstein would not have gone beyond the inheritance of Newton. Special relativity is instead built around two crucial insights, one conceptual and one technical. The former generates a philosophical revolution, the latter a calculational one.

The conceptual insight is that space and time are knit from the same cloth. There is not, contra Newton, any division into absolute space and absolute time, with 'separation in space' and 'separation in time' being two entirely disjoint concepts. In relativity this division is not absolute, but varies from observer to observer. I, stationary in my armchair of reflection, may see two events as simultaneous and with a purely spatial separation. You, passing by on a bicycle – or more realistically, a spaceship travelling close to the speed of light – would instead see one event as clearly earlier than the other.

This insight does not imply that time has no meaning. One of the most important roles of time is to mark causation, and to say what was 'before' and what was 'after'. The sequence of causation remains unaltered in relativity. Some events cause other events and must always precede them. Whoever you are, you will agree that the event 'My son Alexander was born' came later in time than the event 'I was born' – the former could not have occurred without the latter. However, different observers may disagree on whether the event 'I was born' comes before or after the event 'The alien Zak on the planet Zorg was born'.

The second great insight is more technical in nature. It was already known that the laws of physics do not depend on where you are, or which direction you are looking in. Experiments in Madrid, Mumbai and Moscow get the same results, and the laws of physics do not care whether experiments are located in a north-facing building or a south-facing building. There can be local environmental effects: June in Madrid and December in Moscow require different levels of thermal insulation. However, once these environmental effects are removed the underlying laws of physics have no preferred orientation, as the three spatial directions can be rotated into one another without changing the laws of physics.

Einstein also realised that these rotations can be extended to include the time direction. There is a particular set of rotations that mix the space and time coordinates, under which the laws of physics are invariant. This rotation is called a *Lorentz transformation* and underlies all the mathematics of both special and general relativity.

It is not the purpose of this book to give a detailed treatise on relativity or to describe its mathematics. The essence of relativity was neither the mutual identification nor the abolition of space and time, but instead the realisation that they are both cut from the same cloth and are part of a single object: spacetime. In the words of one of the founders of relativity, the German physicist Hermann Minkowski: 'Space of itself and time of itself will disappear into mere shadows, and only a kind of union between the two will survive.'

3.2 SPACETIME IS DYNAMICAL

Special relativity took Newton's vision of absolute space and absolute time and destroyed it. Physics according to Newton involved events in space occurring to the backdrop of a universal time measured on a divine clock. This vision vanished with special relativity, and Newton's concept of absolute time died a complete and irreversible death. As the arena in which the laws of physics played out, space and time were replaced in special relativity by the single object spacetime. This new entity then became the unchangeable background against which the laws of physics are formulated.

This insight belonged to Einstein. It is to his great credit that he also made the further discovery that the new picture is also wrong – special relativity must in turn be replaced by general relativity. General relativity, which Einstein developed in 1915, takes the immutable spacetime of special relativity and breathes life into it. The single most important point of general relativity is that spacetime itself is dynamical. The geometry of spacetime evolves in response to matter. Matter tells geometry how to bend, and geometry tells matter how to move. The fixed spacetime of special relativity is revealed to be no more fixed than the fixed stars of Ptolemaic astronomy.[1]

The details of exactly how geometry responds to matter and how matter responds to geometry lie within the equations of general relativity. Superficially, these equations are highly complex. There are ten of them, and they all depend on each other and have to be treated at the same time. For all but the very simplest cases, they are hard to solve exactly.

However, properly understood, these equations describe one of the most conceptually simple of all scientific ideas.[2] The gravitational response of matter to geometry in the curved spacetime of general relativity is extremely easy to state: matter always continues in the direction it is already going, following the shortest, straightest path it can find. What could be simpler? Aircraft flight paths, which appear as funny curves on two-dimensional projections of the earth, become straight lines when viewed on an actual globe. In curved spacetimes, this principle of shortness and straightness already encompasses all the motion of bodies in Newtonian gravity. Within general relativity, the inverse square force law of Newton's law of gravity is simply a consequence of the statement that bodies travel in straight lines in curved spacetime.

General relativity is the paradigmatic example of good theoretical physics. It is mathematically beautiful and elegant. It is conceptually simple, with no free parameters and no adjustable knobs that need tuning. It is also physically true and gives a precise description of many aspects of the universe. Its scientific virtues are completed by the fact that it is also practically useful, being essential for the correct functioning of GPS devices.

That spacetime is dynamical – and the geometry of the universe itself is evolving – is also the foundation of modern cosmology. As we will discuss further below, observations of distant galaxies require for consistency that as

---

[1] In the classical astronomy of Ptolemy, who lived in the second century AD, the stars were fixed in a celestial sphere that lay far beyond the region of the planets. This sphere rotated daily, producing the movement of the stars through the sky when viewed from the fixed earth.

[2] It is often a mark of the greatest breakthroughs that they are both complex and simple: complex viewed from the old perspective and simple from the new.

time goes on, the space between the galaxies must be increasing. As time goes on, there is simply more space between the galaxies than there used to be. The overall geometry has changed and space itself has stretched.

We give a one-dimensional analogy for this expansion. We consider two small ants on an elastic band. To start with, the band is unstretched, and the ants can walk around the band from one to another. We can imagine them as separated by a particular finite distance – say five centimetres – such that if one ant wanted to walk across to its friend it would take a particular time, for example ten seconds. We now gradually start pulling the band so that it stretches. As we do so, the separation between the ants increases. Neither ant has moved by themselves, as they have in no way exerted themselves. However, as we continue pulling on the band the ants progressively move further and further apart. The amount of time it would take one ant to travel to the other also grows: from ten seconds to thirty seconds to five minutes and more – the intermediate space has grown and expanded, and its geometry has changed.

In this example, the change in the geometry of the band is imposed by an external agent (the person pulling on its ends). The band does not stretch dynamically by itself, but only through applied external forces. In general relativity applied to cosmology, the growth of the universe and the change in its geometry is dynamical – it is determined entirely by the sources of matter and energy within the universe.

The dynamical nature of spacetime is also crucial to the existence of black holes. In normal circumstances the curvature of space is small. Regular matter – the earth, or even the sun – produces only small local distortions on a flat background spacetime. Why do I say small? In physics, 'small' and 'large' are always relative concepts. The sun's gravitational force may be large by human standards, but it is perfectly described by Newton's theory of gravitation, and 'small' in this context means 'well described by Newton'. The sun is neither sufficiently dense nor sufficiently massive to affect spacetime. As an analogue for these distortions, imagine a tennis ball placed on a tight trampoline. The trampoline bends slightly but retains its shape. However, if enough matter is put into a small enough region, these distortions are no longer small. The analogy is no longer a tennis ball on a trampoline, but a beach ball filled with concrete. Spacetime is no longer similar but slightly curved, but bends in on itself beyond the point of return. Given enough matter, spacetime curves so much and so badly that it produces a region from which nothing can escape – a black hole.

Black holes are a consequence of the dynamics of spacetime. Spacetime curves in response to matter, and if enough matter is present, spacetime curves enough to form a black hole. The densities required are large. A black hole with the mass of the earth would require all its matter to be squashed into a region nine millimetres in radius. The 'squash radius' grows linearly with the mass. The sun, one hundred and fifty thousand times heavier than Earth, would need to be compressed into a region around three kilometres in radius – which is indeed the typical size of the black holes that are formed as an endpoint of stellar evolution.

Einstein's theory of general relativity represents the highest development of classical physics. It made subtle modifications to Einstein's previous theory of special relativity, modifications that would require several decades for decisive experimental test.

However, it was only a few years after Einstein had constructed this theory that its classical foundations would be swept away.

3.3   THE WORLD IS DESCRIBED BY QUANTUM MECHANICS

I now turn to what was by a long way the most important discovery of twentieth century science. This is the fact that the world is described by quantum mechanics. No other scientific discovery, either before or since, has combined to the same degree foundational insight, philosophical significance and technological impact – not the discovery of DNA, not the laws of electromagnetism and not even the theory of evolution. A scientific outlook unenlightened by any contact with quantum mechanics is partial and limited. Such an outlook produces a view of nature that is foreshortened and darkened, attached to an understanding haunted by premodern incubi.

The importance of quantum mechanics is threefold. First, it overthrew classical Newtonian mechanics, a theory so successful and influential that the German philosopher Immanuel Kant saw it as an absolute truth, even almost a necessary one. The discovery of quantum mechanics showed that these fundamental laws of dynamics were wrong, both calculationally and conceptually – and not just for objects moving at close to the speed of light. Second, quantum mechanics has through its uncertainty principle permanently modified our ideas of what can be known, and even what it means to know something. Last and by no means least, quantum mechanics has had immense technological application. It is not only the foundation on which all of chemistry is built, but is also essential for the construction of devices ranging from lasers to semiconductors.

Newton's laws provided a deterministic outlook for how matter behaves. In Newton's vision, bodies move because forces act on them. In this world objects have both a position and a velocity. After a small instant of time, the positions have changed as the bodies are moving, and the velocities have changed because there are forces acting. All the quantities one associates with an object – its position, its velocity and its energy – are also continuous quantities that can change smoothly and can take on any value.

This vision is wrong in many ways. The 'quantum' in quantum mechanics refers to the fact that the smoothness no longer holds. Energies can become discrete, taking only specific, fixed values. The possible energies an electron can have within a hydrogen atom cannot change continuously – there is a discrete set of possibilities. These discrete jumps of energy are called quanta of energy. At small distances, positions and velocities cease to be good concepts: an electron in orbit around an atom cannot be said to have a position, but only a probabilistic distribution of possible positions.

Newtonian mechanics also offered not just a dynamics but a philosophy – Newtonianism. This philosophy was admittedly not that of Newton himself, that deeply religious, semi-magical genius. However, the influence of Newton's ideas spread widely. Newtonian mechanics is entirely deterministic. Given the initial values of the positions and velocities of particles, their future evolution is fixed. Arbitrarily precise measurements now lead to arbitrarily precise predictions for the future. If you know the positions and velocities of all particles, all aspects of the future – war, love, death – are predetermined for all time. At the point when the first humans rubbed flints together to make fire, the scorer of the winning goal in the 2018 football World Cup was already set. As Newton's ideas were popularised in France (by among others Voltaire and his mistress Émilie du Châtelet), they produced a vision of a determinist future, where sufficiently precise measurements would allow sufficiently brilliant French mathematicians to calculate the future to arbitrary accuracy. This proud vision was eventually followed by the proverbial fall, as hubris led to nemesis. Newtonian determinism is not the way world works: in quantum mechanics, the uncertainty principle tells us that we can never know positions and velocities simultaneously. It is not just that we do not know; it is that we can *never* know.

Quantum mechanics is also the great organising principle on which chemistry is built. It is empirically true that the different chemical elements can be classified into different types. Some elements are highly reactive and immediately bind into compounds – for example sodium and chlorine. Others are inert and stable, such as the noble gases Neon, Argon and Xenon. This knowledge is summarised in the periodic table, made famous by the Russian chemist Dmitri Mendeleev. By itself, this is an alchemical almanac of mysterious patterns and unexplained principles. The behaviour of elements repeats periodically at intervals, but the sizes of the intervals – 2, 8, 8, 18 ... – are unexplained. There is a structure, but it is unclear where the structure comes from. Quantum mechanics reveals the origin of these apparently mysterious numbers as simply the number of solutions of a particular equation that describes the quantum mechanics of atoms, Legendre's equation, named after the eighteenth-century French mathematician Adrien-Marie Legendre.

Quantum mechanics provides a reason for the rules of chemistry. It explains both qualitatively and quantitatively why different elements have the properties they do, and why certain elements like to bond with other elements. One of the great eureka moments as an undergraduate physicist or chemist comes from solving the fundamental equation of quantum mechanics, called the Schrödinger equation, for the system of a heavy positive charge – a nucleus – and a light negative charge – an electron. Given the right preparation the calculations are not hard, and the result immediately reveals the basic structure of the periodic table.

It is true that only for simple systems can the equations of quantum mechanics be solved exactly. 'Simple' here means atoms with only one electron, while neglecting any effects associated to special relativity. However, a requirement for exact solutions with no approximations – really none at all – is a requirement born in mathematics and not physics. It may not be possible to solve the equations exactly, but with large enough computers they can be solved using approximation methods to any accuracy that is physically useful. The use of such approximation methods has become standard, even if for big problems with big equations, big calculations and big computers may be required.

None of this affects the underlying truth: chemistry and chemical reactions are driven by quantum mechanics. The chemical bonds that bind all molecules from simple diatomic combinations to the largest biological proteins are all products of quantum mechanics, and they arise from the solution of its equations.

The physics of quantum mechanics enters not just the microscopic but also the macroscopic properties of matter. As will be discussed at greater length in chapter 8, the bulk properties of metals and materials are also described by quantum mechanics. The flow of electric current is set by whether a material is a conductor, an insulator or a semiconductor. Metals conduct, and ceramics insulate. The reasons for this originate in the quantum mechanics of electrons in matter. For example, metallic conductance arises from a sea of electrons that are free to transport electric charge throughout the material. Why is this sea of electrons present for metals but not for insulators? It is there because the equations of quantum mechanics say it has to be there.

The discovery of quantum mechanics belongs to the early parts of the 20th century. It took a quarter of century to go from nothing to the full equations of quantum mechanics. The first awakening came with the introduction in 1900 by Max Planck of his eponymous constant, through which he was able to solve a puzzle whereby the classical theory of light predicted that an oven would have infinite energy.[3] Planck's constant formed a key part of the 1905 proposal by Einstein that light energy came in individual packets of energy – quanta – called photons. At first quantum mechanics existed as a mongrel theory, either quassical or clantum – one part classical, one part quantum. The need for a full theory of quantum mechanics was recognised, but the form it took was not known. The correct theory of quantum mechanics was suddenly formulated within a few years in the middle of the 1920s, and the names associated with this formulation – Paul Dirac, Werner Heisenberg, Wolfgang

---

[3]Max Planck's life was one that passed through many ages. Born in 1858 into a prosperous and intellectual German family, Planck had a long and successful career, winning the Nobel Prize in 1918. However, before he died in 1947 he would also live through the death of one son in the first world war, the death of two daughters in childbirth, the collapse of Germany into Nazism, the destruction of all his scientific records and correspondence in an Allied bombing raid and finally the execution of a second son for participation in the attempted assassination of Hitler.

Pauli, Erwin Schrödinger – have entered the scientific pantheon. This was one of the truly great moments in science, where a clear and correct revolution took place over the course of a few years, brushing away results that went back centuries.

All the protagonists except Schrödinger were in their twenties. The Germans called it Knabenphysik – boy physics. Wordsworth's words about the French Revolution can be applied to this period with no dark overtones:

> Bliss was it in that dawn to be alive,
> But to be young was very heaven!

3.4 THE WORLD REALLY, HONESTLY, TRULY IS DESCRIBED BY QUANTUM MECHANICS

The next great insight is that the world really, honestly, truly is described by quantum mechanics. Quantum mechanics is sufficiently strange and sufficiently counterintuitive that this point merits repeating. Looking back, much of the history of 20th century physics consists of smart people trying, ultimately in vain, to deny the implications of quantum mechanics. Time and again, apparently paradoxical results have led to doubt that the formalism of quantum mechanics was correct, when the correct attitude was to spend less time doubting and more time thinking hard about how to understand the theory.

A famous early example of this is the EPR paradox. E, P and R stand for Einstein – even Homer nods! – Podolsky and Rosen, and this gang of three put forward this paradox in 1935. The original formulation involved the spin of electrons, but we will make it more vivid. We consider a double Schrödinger cat experiment featuring Patches and Milky. Both Patches and Milky are shut up in separate boxes. The experiment involves a single atomic nucleus, which has a 50 per cent chance of undergoing radioactive decay within a given time period. The respective fate of the cats is determined by whether (or not) this decay occurs. If the nucleus has decayed, Milky gets cyanide and Patches fresh mousemeat. In the absence of the decay, it is Milky who feasts and Patches who perishes. After the fateful time, there are two possible states of the system. Either Milky is in cat heaven and Patches is alive, or Patches has passed on and Milky is eating. It is a fact that in quantum mechanics, we do not know the state of a system until we measure it. The system is therefore in a superposition of these two cases, and we cannot know which case is true until we look inside one of the boxes and perform a measurement. However, the fate of Milky and the fate of Patches are entwined and not separable. If we see that Milky is alive we know that Patches is dead, and vice-versa. Technically, such a quantum state is said to be *entangled*.

To set up the EPR paradox, we first wait until the fateful time. We then do not look inside, but instead load each of the boxes, both the one containing Milky and the one containing Patches, into separate spaceships. These spaceships are launched and travel several lightyears apart in opposite directions. At this point, a crew member on one spaceship opens Milky's box and looks in. The experiment has been set up so that there are only two states that can possibly apply – either Milky is dead and Patches is alive, or Patches is dead and Milky is alive. The essence of the EPR paradox is that when the crew member looks at Milky, this automatically determines the fate of Patches – who by now is many light years distant. How, EPR asked, can a measurement *here* on Milky immediately affect the state of Patches *there*? This is in apparent contradiction to the basic postulate of relativity that nothing can travel faster than the speed of light. The attitude EPR took was that the paradox implied quantum mechanics was incomplete: the full theory must contain 'hidden variables' that we are unable to see.

On this occasion, Einstein was on the wrong side of the argument. Quantum mechanics is indeed how the world works. A careful analysis shows that there is no conflict with relativity. Information is not sent faster than the speed of light, as the results of the measurement on Milky cannot be immediately communicated to the crew on Patches' spaceship. It is only this that is the key requirement of relativity, and this is not challenged by the paradox.

Furthermore, the 'hidden variables' theories that were thought to be necessary have now been chased into so many foxholes that they are almost entirely ruled out. During the 1960s the CERN physicist John Bell devised a series of tests to distinguish experimentally between the proposed hidden variable theories and the standard formulation of quantum mechanics. These tests were carried out experimentally by a group in Paris led by Alain Aspect throughout the 1970s and 1980s, finding in every case agreement with standard quantum mechanics.

A second reaction against quantum mechanics occurred almost contemporaneously with the EPR paradox. Once the quantum mechanics of individual particles was understood, it was natural to extend this to the quantum mechanics of the fields that exist throughout spacetime – in particular, the electromagnetic field. The equations of electromagnetism were written down and quantised so that they became quantum equations rather classical equations. No sooner had this been done than a serious problem was encountered. Classical electrodynamics – the theory written down by James Clerk Maxwell in the 1860s – works. It describes electromagnetic effects well in almost all circumstances. Given it works so well, quantum corrections should be tiny in almost all circumstances. They were not. They were not even small, or even close to small. Every time the Nobel Prize-winning physicists of the 1930s attempted to calculate quantum effects in electromagnetism, they got, instead of a small number, infinity. The quantum 'corrections' entirely dominated the original classical result. Something had clearly gone badly wrong, and that something had arisen in the quantum mechanical calculation.

Faced with these apparently meaningless results, a natural reaction was to discard the formalism. There appeared to be something sick with quantum mechanics applied to fields such as electromagnetism. Quantum mechanics was a truly weird theory, and there was still a residual suspicion of it. Perhaps quantum mechanics needed to be reformulated, or perhaps another theory altogether was necessary. Quantum mechanics itself had been so radical it did not seem unreasonable that another radical idea would be necessary again. The 1965 Nobel Laureate Julian Schwinger said of this period,

> The preoccupation of the majority of involved physicists was not with analysing and carefully applying the known relativistic theory of coupled electrons and electromagnetic fields but with changing it.

Even Paul Dirac, one of the original founders of quantum mechanics, went as far as to propose modifying quantum mechanics through the introduction of negative probability, whatever that might be.

Confusion existed for a long time before physicists eventually realised that they already had the correct equations and the correct formalism. The necessary priorities were first, to learn how to calculate correctly with the infinities, and second, to understand them.

The string-and-sealing-wax approach to dealing with the infinities was separately developed in the 1940s by Richard Feynman, Julian Schwinger and Sin-Itiro Tomonaga. The three were very different. Feynman was lively, informal, charismatic and intuitive. He played the bongo drums and frequented a strip club. Schwinger was a former child prodigy with immense calculational prowess: it would be said of Schwinger that other physicists gave talks to show how to do a calculation, but Schwinger gave talks to show that only he could do it. His technical excellence extended to his lectures, which were virtuoso models of clarity and preparation, but as a person he was private and reserved. Tomonaga was from Japan, descended from the samurai class and born to a father who was a professor of western philosophy. He had been inspired to study physics after hearing Einstein lecture in Kyoto in the 1920s, and his Nobel Prize-winning research was done in the isolation of wartime Japan.

Together, their achievement was to understand how to reduce the plethora of infinities to a set of finite answers. Their approach established the theory of quantum electrodynamics with a set of defined calculational rules, and they were subsequently jointly awarded the 1965 Nobel Prize. The methods used were not entirely pretty. These methods acquired and have not yet shed a slightly inglorious reputation for systematically adding and subtracting infinities so that the eventual answer was finite – 'Just because something is infinite does not mean it is zero'. This can lead to mumblings among the uninitiated about whether quantum field theory really is deep science or just a black magic recipe from a witches' cookbook.

The systematic understanding was provided by Kenneth Wilson in the 1970s. However, first let me try and unpackage the older techniques a little bit. It is true that, whatever quantity in quantum electrodynamics you try to compute, you end up with infinity. However, there is a precise sense in which the sources of these infinities are the same each time. To illustrate this, imagine trying to compute three separate quantities, and getting first

$$\frac{1}{3} + (1 + 2 + 3 + 4 + 5 + \ldots),$$

next

$$\frac{1}{6} + (1 + 2 + 3 + 4 + 5 + \ldots),$$

and finally

$$\frac{1}{7} + (1 + 2 + 3 + 4 + 5 + \ldots).$$

Each sum may not be meaningful by itself, but the differences certainly are meaningful. The difference of the first two sums is one sixth, and that of the second and third sums $\frac{1}{42}$.

The methodology is that of isolation and elimination. First, carefully identify the 'same infinity' that crops up each time, and systematically subtract it to get finite answers. While the absolute value of any single quantity cannot be determined, the difference of two quantities can be predicted. This approach works, and it underlies the spectacularly successful predictions of quantum electrodynamics.[4] However, it does not really explain *why* this method works and it does not give a physical understanding of why the infinities appeared in the first place: at best, they are ugly, and at worst they cast suspicion on the whole approach.

As just mentioned, this understanding was supplied by Kenneth Wilson in the 1970s. In broad terms, Wilson's contribution was to explain where the infinities really came from. They arose from assuming that a theory such as quantum electrodynamics can hold down to arbitrarily small distance scales. It is familiar in physics that equations have a finite realm of validity and break down outside it. For example, Newtonian mechanics fails both when objects move close to the speed of light and also at distances comparable to the size of atoms. Quantum electrodynamics might be correct down to distances of a nanometre, a picometre or a femtometre – but this does not mean it will work for all length scales, however small. In fact, we know that it should not work at all length scales. To take the most extreme example, at some point the quantum effects of the gravitational force will become important. While we might not know exactly what will happen there, it is a fair bet that our existing theories will be modified. The way to get infinite answers is by making the naive assumption that the existing theory holds entirely unmodified over the twenty-one powers of ten in length scale down to the quantum gravity scale – and then also all the way beyond that.[5]

---

[4]Technically, there can be more than just one source of infinities: however, this method can be used as long as the number of sources of infinity is finite.

[5]Even the greats fossilise. Paul Dirac, although one of the founders of quantum mechanics and one of the pioneers of quantum field theory, could never truly accept quantum field theory due to the infinities that arise. Already at retirement age when Wilson did his pioneering work, he was never able to internalise Wilson's conceptual insights.

Wilson's work also offered an additional technical insight (these following paragraphs can be freely skipped if desired). Even if this understanding removes the infinities by explaining why we expect some new physics to come in at some incredibly short length scale, it still appears to leave the puzzle of why measured quantities do not still end up finite, but far beyond the known reach of experiments. If we compute the mass of the electron in quantum field theory, we do get an infinite answer. However, in practice the measured mass-energy of the electron is smaller than the expected energy of quantum gravity effects by a factor of 10^{21}. Even if the infinite corrections are capped at some high scale to give finite answers, why are these still not so large that they drag the electron mass far above its measured value?

The answer is that the infinite corrections involve the logarithm of the high scale, multiplied by the electromagnetic fine structure constant – which is around $\frac{1}{137}$. Even with the high scale set at the quantum gravity scale, the numerical size of the correction is small.[6] Wilson was therefore both able to explain why the calculational infinities were not physical, and also why physics at low energies is only minimally sensitive to physics at high energies.

With or without Wilson's understanding of the infinities, quantum electrodynamics was a triumph. It gave accurate predictions for quantum effects in electromagnetism – in some cases, accurate down to one part in ten billion. The moral to draw from this success should have been that quantum field theory worked, that it was the correct framework to describe particle interactions, and that more time needed to be spent understanding it.

This is the moral that should have been drawn. In the modern telling, all the non-gravitational forces – the strong force, the weak force and the electromagnetic force – are indeed described by quantum field theories. That of the electromagnetic force is the simplest, but they are all just different examples of quantum field theories. In the 1960s, the view looked different. The weak interactions did not satisfy what was believed to be a basic tenet of quantum field theory. Technically, they were not *renormalisable* – a property that was believed to be essential to allow any description in terms of a quantum field theory. Theories that were not renormalisable appear to have an infinite number of infinities in them, which would invalidate the technique of extracting the 'same infinity' and systematically subtracting it off each time.

The experimental situation with the strong force was also mysterious, as an abundant harvest of particles upon particles upon particles appeared. Neither force looked anything like quantum electrodynamics. This, and the quantum field theory formalism that accompanied it, was regarded as a special and

---

[6] A quantitative estimate of the correction to the electron mass is

$$\text{Correction} = (\text{Original mass}) \times \frac{1}{4\pi} \times \left(\frac{1}{137}\right) \times \log\left(\frac{\text{High scale}}{\text{Original mass}}\right),$$

which is numerically small compared to the original electron mass, even when the ratio within the logarithm is 10^{21}.

irrelevant case. A new – perhaps radically different – approach was felt by many to be needed to deal with both the strong and electroweak interactions.

Their will was weak, and their understanding dark – for the end result was, yet again, the triumph of quantum mechanics and quantum field theory. As the Harvard theorist Sidney Coleman put it,[7]

> [1966 to 1979] was a great time to be a high-energy theorist, the period of the famous triumph of quantum field theory. And what a triumph it was, in the old sense of the word: a glorious victory parade, full of wonderful things brought back from far places to make the spectator gasp with awe and laugh with joy.

Those who invested time in the 1960s working on understanding quantum field theory obtained a return that was the scientific equivalent of placing savings in Berkshire Hathaway shares.[8] This can be illustrated through the citation record of one of the foundational papers of the Standard Model, Steven Weinberg's paper 'A Model of Leptons', which was published in 1967.[9] This measures the number of times a paper is referred to by a different paper.

Citations in 1968:	1
Citations in 1969:	1
Citations in 1970:	1
Citations in 1971:	5
Citations in 1972:	36
Citations in 1973:	48
Citations in 1974:	113
Citations in total:	> 9000

The strong and weak interactions are simply further examples of quantum field theories, the logical consequence of the combination of quantum mechanics and special relativity. Despite the best attempts of the best physicists, quantum mechanics really works, and it really does describe the world.

---

[7]Coleman was famous, among other things, for the introductory lectures on quantum field theory he gave at Harvard for many years during the 1970s and 1980s. These lectures were filmed and are now available online. To the modern viewer one of the most striking features of these lectures is the cigarette attached to one of Coleman's hands in the same way that the chalk is attached to the other.

[8]My recently retired Oxford colleague and fellow of the Royal Society, Graham Ross, launched a distinguished career in the 1960s by the simple expedient of studying and learning quantum field theory when almost no one else thought it worthwhile.

[9]Weinberg, who alongside his research has written with grace and learning on a variety of topics, is one of the leading scientific intellectuals of the last hundred years. In addition to research that produced a Nobel Prize, he is the author of many books, both for the general reader and for professional scientists. Mostly right and always interesting, when he speaks the wise attend.

3.5 NATURE LIKES SYMMETRIES

If the kaleidoscopic progress of all of 20th century physics had to be summarised by a single word, that word would be symmetry. More than any other single idea, the identification and applications of symmetries have connected the progress made in apparently disparate areas. These disparate areas include both special and general relativity, high energy particle physics and many-body phenomena such as superconductivity.

The basic concept of a symmetry is simple and familiar. You take an object, make a change, and find that the object still looks the same. For example, a square rotated by ninety degrees looks precisely the same as it did before the rotation. Alternatively, take a hexagon, rotate it this time by sixty degrees, and the same hexagon looks back at you. The form of the symmetry is different – both a hexagon rotated by ninety degrees and a square rotated by sixty degrees do change their shape. Both the square and the hexagon exhibit simple geometric rotational symmetries, and are said to have respectively fourfold or sixfold symmetry.

Something similar, but subtly and slightly different, happens with the circle. Given a perfect circle, you can rotate it by a tiny amount and find that you still have the same circle. However, this time I do not need to be precise about what a 'tiny amount' means. For the square, the rotation needed to be through ninety degrees exactly, and for a hexagon, sixty degrees exactly. For the circle, no precision is needed. Rotations by either one-tenth of a degree or 17.3456 degrees return the circle back to itself. Instead of simply a few discrete choices of rotation angle preserving the shape, there now exists a continuous infinity of choices.

As 'number of degrees we can rotate by' is a continuous quantity for the circle, but one of a discrete set of choices for the square or hexagon, the symmetries are named differently. The symmetry of the circle is said to be a continuous symmetry and the symmetry of the square is said to be a discrete symmetry. Both types of symmetry, continuous and discrete, have played important roles in physics.[10] The study of symmetries in and of themselves is a fine topic for mathematicians. The proof of one mathematical result related to symmetries, the so-called classification of finite groups, is reckoned as the longest published proof in all mathematics, coming in at around ten thousand pages once all contributing journal articles are included. We shall return to one aspect of this result in chapter 9. However, to a physicist a symmetry is bereft of relevance unless it has something it can usefully apply itself to, in the way that the rotational symmetries above could act on the square, the hexagon and the circle.

Given that symmetries have been so important in physics, what then is their target? The target is nothing more nor less than the laws of nature themselves. The symmetries apply to the fundamental laws of nature. There

---

[10]As we shall discuss further below, for the case of discrete symmetries, symmetries-that-are-almost-but-not-quite-symmetries have perhaps been most useful.

are certain rotations or transformations that we can perform on these laws but that nonetheless leave them unaltered. Two different observers related by symmetry both observe the same laws of physics.

For example, the laws of physics have been discovered to possess rotational symmetry. Suppose I plant three wooden beams down, all at right angles, using these to define three axes x, y and z. I measure the laws of physics – for example, the statement that force equals mass times acceleration – with respect to these axes and coordinates x, y and z. I now rotate the frame about the z-direction, such that the x-direction points in the y-direction and the y-direction points in the minus-x direction, and repeat the measurements. I will get the same results and find the same laws of physics. The fundamental laws of physics are unaltered when expressed in terms of these new coordinates.

This result is an amazing discovery. It should come across as shocking and outrageous, a mixture of a 'what?' and a 'wow!'. If it does not, it is a result of both the complacency of hindsight and our inability to recast our minds to an earlier way of thinking. Why is this result so shocking? For rotation in a plane, the result is indeed plausible. We do not fret about whether athletics records should be recorded differently for tracks laid out north-south or east-west. Our ability to walk, run and hunt is unaffected by the direction in which we are facing. We tell the difference between east and west by the sun, not by noticing differences in how objects move. However, this is not at all the case for three-dimensional rotations. In fact, the lessons of physical experience always and everywhere contradict any symmetry between up and across – and anyone doubting this is invited to try and climb a tree and then fall sideways. Gravity, the force that we first meet in the cradle, appears to act only in the vertical direction, and then specifically down rather than up. This singles out both a preferred axis (the up-down one) and a preferred direction (down).

These results appear so obviously true that they threw natural philosophers, the proto-scientists of their day, for millennia, and an entire false physics was constructed around it. This was the Aristotelian universe, radiating out from the centre of the earth. Everything is made of four elements – earth, air, fire and water – each of which seeks to find their natural location. The centre of the earth is the centre of the universe, and in that objects partake of the element 'earth' they want to fall inwards towards this point, whereas the element 'fire' wants to rise to the heavens. Aristotelian physics gives an intuitive picture correctly describing our day-to-day experience – and is also totally and irrevocably wrong.

For, whatever our daily experience may say, the fundamental laws of physics are indeed symmetric under three-dimensional rotations. Gravity has no intrinsic preferred direction. It simply attracts stuff to other stuff, causing objects to clump together. If you put a thousand billion billion tons of stuff – let us call this the earth – down somewhere, by doing so you introduce a preferred direction for everything else towards this great pile of matter. However, this preferred direction is only accidental. Moths flying in a dark room have no preferred direction of flight. Light a candle, though, and they will cluster

around it. In a similar way, the apparent preference towards the downwards
direction on earth is real but accidental, set by the presence of the earth but
not part of the basic rules of the game.

Even though it alone should be enough to shock, rotational symmetry is
only the simplest of the symmetries that apply to the laws of physics. It is also
a symmetry that had been recognised prior to the twentieth century. Much
of the progress in physics during the last century has come from a broader
development of both the types and applications of symmetries.

As with space, so with time. Special relativity comes from extending the
rotational symmetry of space to include time. Instead of simply considering
rotations among three spatial coordinates x, y and z, rotations now involve
both space and time, involving also a fourth coordinate: time t. In spatial
rotations, the roles of x, y and z in equations would be mixed up and interchanged, but leaving the equations with ultimately the same form. In the
spatiotemporal rotations of special relativity, the roles of x, y, z and t in equations are interchanged: but as before, the requirement of symmetry is that the
laws of physics take the same form both before and after the rotation. This
requirement enforces a special form for the physical laws – the structure of
special relativity. While this discovery of Einstein is sometimes introduced by
thinking about what it would be like to travel near the speed of light, it can
also be viewed as the inevitable result of symmetry under spatiotemporal,
rather than just spatial, rotations.[11]

General relativity is built on a cognate but more audacious use of symmetry. Special relativity makes the laws of physics invariant under some particular choice of coordinates. We set down our wooden spatiotemporal measuring
frame, and however we rotate it, the laws of physics will still take the same
form. Special relativity does however require the presence of the frame. As one
example, the spatial legs of the frame must be at angles of ninety degrees. We
can rotate the frame how we wish, but we cannot choose its legs arbitrarily.

General relativity does away with this. The essence of general relativity
is that the laws of physics do not care for coordinates. Coordinates in all
their incarnations – atlases, charts, grid references – are mere human con-
structs, and nature does not give a fig about our choice of coordinate systems.
Does the earth know about latitude and longitude? Did the great white whale
inwardly rejoice as it crossed the prime meridian? The physical laws are ultimately equally indifferent to all choices of coordinates. Einstein's greatest
insight was to require that the laws of physics be independent of coordinates.
Whichever choice of coordinates you lay down, the laws of physics should take
the same form. The lift from special to general relativity comes from demanding the symmetry of rigid rotations be extended to any choice of coordinates

---

[11]The mathematics of these rotations involves one subtlety. The time part of these rotations effectively involves an extra minus sign compared to purely spatial ones. As $i^2 = -1$,
the mathematical fiction of 'imaginary time' renders the mathematics identical to that of
spatial rotations, at the cost of sounding entirely baffling.

whatsoever. While this may sound deceptively simple, it is a measure of Einstein's intellectual depth that this insight also leads to so many consequences.

This symmetry – whose technical name is *general covariance* – marks the high point of classical physics. Formulated in 1915 amidst one of the low points in European history, general relativity describes the dynamics of space and time themselves. It is the crowning glory of classical physics – but symmetry principles would be no less relevant for the quantum mechanics that would follow a few years later.

The symmetries discussed so far have been geometric symmetries, building on intuitions developed for two-dimensional rotations. Their implications may be hard to understand, but their roots are clear. However, there is another type of symmetry that is less easy to intuit, called *gauge symmetry*. These gauge symmetries are to the Standard Model of particle physics what carbon is to organic molecules: the foundations on which the entire structure is built. While incredibly important, they are also harder to explain than geometric symmetries, as their symmetries are internal to the mathematical equations of the Standard Model rather than inherited from the outside world.

Gauge symmetries are not easy. One of the many remarks alleged to have been said by Einstein – like Mark Twain, he attracts quotations – is that if you understand something, you can explain it to your grandmother. The logical consequence of this would be that either Granny Einstein was a mathematical whizz, I do not understand gauge symmetry, or Einstein understood gauge theory a whole lot better than I do. It requires no attack of hubris or egomania to say that none of the above are true. Science marches on. In the same manner that any rational person in need of an appendectomy would choose the most ham-fisted product of an NHS teaching hospital in preference to Avicenna, so the least prepared of today's physics graduate students far surpass Einstein in their understanding of quantum field theory.

This declaration that gauge symmetry is hard is a prelude to my attempt at explaining it. The following page is however inessential to the narrative and the explanatory attempt can be skipped without harm.

First of all, what is it not? Gauge symmetry is not a spatiotemporal symmetry. It is an internal symmetry of equations rather than an external symmetry such as rotations. What is meant by an internal symmetry? The equation

$$q^2 + p^2 = r^2$$

has an internal discrete symmetry whereby q and p are exchanged. Doing so simply gives back the same equation. Furthermore, this equation in fact has a continuous family of symmetries. To see this, we realise that this is the equation of a circle, with q and p treated as the coordinates and r as the radius. There is then an internal family of rotations by which q and p are rotated into one another – although we should think of these rotations as purely mathematical, acting on the objects inside the equations.

The first requirement for a gauge symmetry is a set of equations that represent physical laws and must hold at every point in space and at every moment in time. An example of these are the laws of electromagnetism. The equations that describe the electric and magnetic fields hold here, there and everywhere; and they hold in the past, present and future. Wherever you are in the universe, these equations hold. What is needed to call these equations a gauge theory is the additional fact that these equations always have, at every instant in space and time, an additional internal rotational symmetry. A theory is then said to possess gauge symmetry if we are able to perform an internal rotation of the equations at every point in space and time, also choosing the rotation to be different at every point (ninety degrees here, ninety-one degrees here, ninety-two degrees here), such that the overall equations still remain unaltered.

This is the most basic idea of what gauge symmetry is – why is it called so? In the rest of the world, gauges appear most prominently in railway layouts. Here the gauge specifies the separation between the rails, and thereby the spacing between the wheels of the train. Different choices of gauge have brutally different effects, as any attempt to run a train straight from a broad gauge railway onto a narrow gauge railway would demonstrate. In physics, the gauge refers to the degree of internal rotation – and gauge symmetry is the statement that both the equations and the physics remain the same as you alter the gauge so that it constantly changes from point to point.

This may indeed seem opaque and complicated. However, you should know – you must know – you will know – that the reason gauge symmetries are so important is because they correspond to forces. Electromagnetism is described by a gauge theory. The strong nuclear force is described by a gauge symmetry. The weak nuclear force is described by a gauge symmetry. Three of the four fundamental forces in nature come down to the special type of symmetry known as a gauge symmetry. It is for this reason that gauge theories play the role in the education of aspirant young particle theorists that the *Iliad* and *Odyssey* played in the classical educations of an earlier era.

It is a further remarkable statement that the gauge symmetries dictate not just the forces that are present but also the types of particle that are allowed and the form their interactions must take. One of the beauties of gauge symmetries, which also makes them highly testable, is the way they force many interactions that appear independent into a particular structure. In the same way that the basic unit of light – the basic unit of electromagnetism – is the photon, the basic unit of the strong interactions is called the gluon. The gauge symmetry of the strong interaction tells us that there must be not one but a total of eight gluons, and it furthermore tells us how each of these eight gluons interacts with each of the others. As another example, the W^- particle decays to the electron, muon and tau particles with almost equal rates. This equality follows because of gauge symmetry, and could not have been otherwise. The gauge symmetry of the weak interaction also relates the interactions of the electron to the interactions of the neutrino.

Why is this? It comes down to what gauge symmetries actually do to the equations. For example, in the equations of the Standard Model there are terms that involve both the electron and the neutrino.[12] From these equations, one can extract the interactions of the electron and also the interactions of the neutrino. The way the gauge symmetry of the weak interaction works is that it rotates the term corresponding to the electron into the term corresponding to the neutrino. Acting on the innards of the equations, it moves the terms around so that the 'electron' term moves to where the 'neutrino' term was, and vice-versa. The gauge symmetry is a symmetry, so after this change the equations of the weak interaction must still take the same form – which they do, except what used to be called the 'electron' is now called the 'neutrino'. In this way the weak interactions of the neutrino are fixed by the weak interactions of the electron.[13]

Let me also try and explain why gauge symmetries fix the numbers of different types of particles. I have just said that one of the main ways gauge symmetries act on the equations is by moving around the terms corresponding to different particles. Different symmetries do this in different ways. I mentioned earlier the case of the fourfold and sixfold rotational symmetries that belong respectively to the square and the hexagon. Both symmetries return to their starting point, but after producing a different number of 'copies': four and six. The number of such copies is set by the mathematics of how the symmetry works. Although the algebra is more complicated, the same principle holds for the gauge symmetries of the Standard Model. The mathematics of symmetry imply that there are only a certain number of independent ways of reordering the particle terms in the equations among each other. The number of such ways that exist tell you the number of each type of particle, all of whose interactions are related to each other by symmetry. It is this mathematics of symmetry that tells us that there are three copies of each quark and eight copies of the gluon.[14]

Everything said above involves deep, powerful and true statements. However the mathematics of this is rather abstract and I am not going to elaborate further. Instead, having considered geometric and gauge symmetries, I want to consider a final topic: symmetries that are almost but not quite symmetries.

This refers to the fact that there exist certain discrete operations that are possible, and under which the laws of physics must clearly and obviously not change – were it not for the fact that experiment tells us they do indeed change. The best example of these is called parity. Parity is the operation that interchanges left and right. The distinction between left and right is a conventional one. I drive on the left in the United Kingdom but in the United States people drive on the right. This is not a reflection of deep cultural or

---

[12]For full technical accuracy, the statements here hold for what is called the *left-handed electron*.

[13]But how can this be a symmetry if the electron has charge and a neutrino has no charge?! This is why the statements in the paragraph only refer to the behaviour under the gauge symmetry of the weak interaction, and not to electromagnetic interactions.

[14]Mathematically, this is called the *dimension of the representation of the symmetry*.

linguistic differences. A choice had to be made; no choice was intrinsically preferable; different choices were made in different places.

The distinction between left and right is so obviously conventional that it is striking that nature, in the form of the laws of physics, does actually distinguish between left and right. What does this mean? Write down the laws of physics in one particular coordinate system, using coordinates x, y and z. Now exchange x with 'minus x', thereby exchanging the meaning of left and right.[15] The laws of physics change form, and the equations take a different form. After the operation, the old laws, applied in the new coordinate system, do not work: they make definite predictions for experimental measurements, and these predictions are wrong.

For a long time, there was no awareness that this could be possible. Experience with electromagnetism did not help as parity is a good symmetry of electromagnetism. The possibility that parity symmetry may be violated in nature was realised as a theoretical possibility in 1956 by the Sino-American pair Tsung-Dao Lee and Chen-Ning Yang. Their suggestion was followed by rapid experimental confirmation by Chien-Shiung Wu in 1957. The observation of parity violation came as a shock, and it led to the almost immediate award of the Nobel Prize to Lee and Yang. The fact that parity is *not* a symmetry of nature was one of the first clues to understanding the full form of the weak interactions. In contrast, the laws of the gravitational, strong and electromagnetic interactions do respect parity, and they take the same form under the interchange of left and right.

The classification and application of symmetries has then played several major roles in twentieth century physics. Geometric symmetries underlie both special and general relativity. Particular 'internal' symmetries, called gauge symmetries, set the nature of the strong, weak and electromagnetic forces. Discrete almost-symmetries, such as parity, determine the finer structure of particle interactions.

We now move from the minutiae of particles to the largest scales in physics.

3.6 THE UNIVERSE WAS YOUNG, SMOOTH AND HOT, BUT IS NOW OLD, COLD AND WRINKLY

The heavens, it was thought, endure from everlasting to everlasting. So it seemed to our ancestors, and so it seems to us. So it also is, on any timescale that we either live by or are biologically sensitive to. Our universe does not change on the timescale of old people – a hundred years – or old civilisations – a thousand years. If Ur the Cro-Magnon had developed general relativity and satellite astronomy, he would have measured the same universe, with the

---

[15]To see that this does exchange left and right, do the following. Curl your right hand in a ball, and then extend your thumb, forefinger and middle finger to define a set of axes. Do the same with your left hand. Your fingers differ only by the exchange of x with 'minus x', but brief contortional attempts will convince you that there is no way to rotate the two axes into another without a broken finger.

same parameters, as ourselves. Even restricting to the Milky Way, over the last hundred thousand years the sun would have moved through only one part in two thousand of its orbit around the galactic centre.

However, on very long timescales we do now know that the universe has in fact aged. We may not know its future, but we do know its past. It has an age, and its complexion and appearance have changed with time. The universe used to be younger. The universe used to be smoother, and the universe used to be hotter.

What do we know about the universe? First, it is not just big, but Big. We – the collective noun for all those who have ever reflected on our place in the cosmos – have always known this. It took us a long time though to realise just how large the visible universe is. Let us allow ourselves to race through history to the point where we can assume the earth goes round the sun at a distance of ninety-three million miles, or a hundred and fifty million kilometres.[16] This is a great distance, but it is small compared to what is to come. We look at the night sky and see stars. How far away are they? How can we work out this distance?

One clue is that the stars do not appear to change their relative position throughout the course of the year, even though during this period the earth has changed its position relative to the sun by around two hundred million miles. Perspective normally depends on our locus of view, and is only unaltered for objects far off in the distance. The apparent positions of the apple tree and the rose bush differ when viewed from the living room or the garden shed – but the distant mountains look exactly the same. Between winter and summer, our viewing platform for the stars moves by a distance of two hundred million miles, yet the relative arrangement of the constellations remains the same. For this to happen, the stars must be as distant from our solar system as the mountains are from the garden.

For this reason, the stars were historically called *fixed* – thus contrasting them with the wandering stars, the planets. Despite the name, the fixed stars are not truly fixed. The stellar positions do change, ever so slightly, when viewed in summer and in winter. The change is not large. It is at the level of one thousandth of a degree or smaller, roughly equivalent to the width of your finger viewed from several miles away. This change is, however, measurable, and is called *parallax*. Once it is measured, trigonometry and knowledge of the earth-sun distance then determine the distance to the given star. Proportionately closer objects have proportionately larger parallaxes. Parallax gives its name to the distance unit of a parsec: one parsec is the distance from earth of an object with a parallax of exactly one arc-second, where one arc-second is three-thousand-six-hundredth of a degree.

These distances are clearly large. It is apparent that such minuscule angular changes, when viewed from points separated by two hundred million miles,

---

[16]We are in no position to sneer at those who thought otherwise. It is less than a century since professional astronomers were all underestimating the size of the visible universe by a factor of ten million.

can only hold if stars are at enormous distances. To measure these distances, the light units introduced in chapter 2 are used. The distance from London to New York is one-sixtieth of a light second, the distance from the earth to the moon is around one light second, the distance from us to the sun is eight light minutes, and the distance to the nearest star Proxima Centauri is four light years – or twenty million million miles. Indeed one parsec equals three and a quarter light years, and an object whose apparent angle changes from summer to winter by only one thousandth of a degree is at a distance of ten light years from earth.

Using these techniques, the distances to nearby stars can be measured, up to a range of approximately one thousand light years. Beyond that, stars are so far away that their positions remain fixed even to our most sensitive telescopes. Parallax is still only the first rung on the cosmic distance ladder. The progression through the rungs to measure distances to objects further away requires several clever tricks. This book is not the place for the full details, but all of these tricks rely on the notion of a Standard Candle.

What is a Standard Candle? You go to the supermarket to purchase twelve identical sixty watt bulbs. Plugged in to sockets, each emits a fixed amount of energy per second. Every second, a hand placed five centimetres, or ten centimetres, or one metre away from the bulb will receive a definite amount of energy. The amount of energy received is determined only by the wattage of the bulb and the distance of the hand from it. If we know the distance, we can work out the energy the hand will receive.

We can then turn this around. By measuring the energy received each second, we can determine our distance from the bulb. For a supply of identical bulbs, it is not even necessary to know the wattage. We take one bulb, find a metre rule and lightmeter, and plug the bulb into a socket. Measuring at a known distance, we obtain a calibrated graph of intensity against distance. Once this is done, we can discard the metre rule: our graph now provides the means of measurement. Using this graph, we can measure the distance to any other Standard Candle bulb purely by measuring the intensity of the arriving light.

The cosmic distance ladder is climbed by reaching from one Standard Candle to another. The only difference is that the Standard Candles used are provided not by the supermarket but by the universe. Starting with nearby objects, the overlapping distance ranges of each Standard Candle eventually take us out to the furthest ends of the universe.

The first Standard Candles are numerous and nearby. Their intensities are calibrated against direct distance measurements via parallax. Using this calibration, the measurable distances can be extended outwards, which in turn allow the determination of the distances to the next Standard Candles, and so on in a chain which spreads out across the visible universe.

The Standard Candles come in various forms. At short distances dim but numerous sources, such as Cepheid variable stars, can be used. At the largest distances, bright but rare objects, such as type IA supernovae, are needed.

In all cases the Standard Candles are believed to burn with a fixed intrinsic brightness: by measuring the light received, we can measure the distance to the Standard Candle. By measuring the distance to the Standard Candle, we can also measure distances to everything close to it – for example, to all other stars within the same galaxy. The upshot is a network of distance measurements stretching out across the universe.[17]

These distance measurements reveal something striking. Everything is moving away from us, and the further away objects are, the more rapidly they are receding. As we go forward in time, the distance separating us from any distant galaxy is continuously increasing. Looking around, we see in all directions the cosmos withdrawing from us.

How do we tell? How do we know whether a faraway galaxy, a billion light years away, is approaching or receding? We cannot, after all, visit it. We tell the same way we tell that a distant ambulance is approaching or receding. An ambulance siren emits a note at a fixed intrinsic pitch, and we are alerted to the ambulance by the siren. The pitch we hear, however, depends on the motion of the ambulance. If the ambulance is coming towards us, the pitch increases as the signals pile up with the added velocity of the ambulance. As the ambulance pulls away, the pitch falls as the gap between notes reaching us increases. Even without seeing it, we can tell from the pitch of its note whether the ambulance approaches or departs.

The same principle applies for distant galaxies. The cosmic siren comes not from sound, which cannot travel through the vacuum of deep space, but from light. Quantum transitions between the atomic energy levels of hydrogen emit light at universal frequencies set by the laws of quantum mechanics. This light shines out across the cosmos, eventually reaching us and our telescopes. As the light originates from particular known transitions within the hydrogen atom, its intrinsic frequency is set. If the arriving light is at a higher frequency than the intrinsic one, the emitting object is moving towards us. If the arriving light is at a lower frequency, the emitting object is moving away.

It is a hard empirical fact that these sirens of distant galaxies, reaching earth, are all received at lower frequencies than the intrinsic frequency, and the more distant the original galaxy, the lower the arrival frequency. The consequence is that all faraway galaxies are receding, and the further they are, the more rapid the recession.[18]

The most distant objects we see are currently separated from us by around forty-five billion light years – and growing. If they are moving away into the future, then in the past they must have been closer. The equations of general relativity can be used to wind the clock back. As we do so, the distance to

---

[17]'Measurement began our might': any error in calibration or in object identification can throw this method into disarray. In the 1950s extragalactic distances suddenly doubled as the German astronomer Walter Baade realised that Cepheid variable stars came in two kinds, and the brighter and dimmer kinds were being confused in distance calibrations.

[18]This fact – recession velocity grows linearly with distance – is a relationship discovered by the American astronomer Edwin Hubble in 1927 and is known as Hubble's law.

far-off galaxies steadily reduces. Going back in time, as objects move together the universe becomes more crowded and more dense. All that we can see today was in the past confined to a smaller volume. Extrapolating the equations naively, we find that approximately fifteen billion years ago everything was on top of everything else. The final few seconds of this extrapolation are conventionally called the Big Bang.[19] According to Einstein's equations of general relativity, as we return to the point of 'zero time' the density of matter becomes larger and larger and eventually, according to the mathematics, infinite at time zero. This infinity implies either the mystical attainment of the eternal Tao of nothingness or that the equations of general relativity have moved beyond their regime of validity. It is unclear precisely when this would occur – one millionth of a second after 'zero time', a thousandth of a second after 'zero time' – but it is clear that the early universe was both far smaller and far denser than the present one.

This contraction has another effect. It is true but not obvious that the universe itself has a temperature. This is not a trick sentence based on double meanings. It means precisely what it says. If you go out into deep space, far away from any stars or galaxies, and place a thermometer there, it will measure a temperature. That temperature is minus 271 degrees Celsius, or 2.73 degrees above absolute zero. Such a temperature is a little parky, but there are places colder: for example the magnets inside the Large Hadron Collider.

The early universe was a crowded place. While the current universe is a bit chilly, as we go back in time, all the light, all the energy and all the matter we see today was squashed into a smaller and smaller volume. The result of squashing so much energy into so small a region is to make the universe hotter. The further we go back in time, the hotter the temperature of this oven becomes. When the universe was a tenth of the current size, its temperature was ten times its present value, and when the universe was a hundredth of its current size, the temperature was a hundred times the present value – or approximately room temperature. When the universe was one second old, its temperature was ten billion degrees.

The early universe was not just much hotter than the present one. It was also much smoother. A defining feature of the cosmos is that it is clumpy. Under the action of gravity, matter has clumped together to form structured objects such as the earth or the galaxy. Matter is not distributed equally: the density of the earth is much denser than the space touching onto the upper atmosphere, which is in turn much denser than at a typical place in the galaxy.

This clumpiness is visible on many different scales. Tiny grains of dust came together to form planets and stars. Planets and stars make up solar systems. The different solar systems are grouped into galaxies, which in turn

---

[19]The originally derisive description 'Big Bang' was coined by the British and Yorkshire astrophysicist Fred Hoyle, who regarded it as an ugly and theistic way for the universe to start. Blunt, brilliant and stubborn, Hoyle got himself into many arguments over the years. His suspicions would not have been allayed by the fact that the first proposer of the idea of a big bang was the Belgian priest-astronomer Georges Lemaître.

join with other galaxies into galactic clusters and superclusters. Each of these groupings looks like a concentration of matter surrounded by emptiness. The earth is much denser than the solar system, and the solar system is much denser than the galaxy as a whole. However, even a spot halfway between here and the nearest star, a cold and inhospitable place if ever there was one, is busy and bustling compared to true deep space. True, deep, empty space cannot be found near this or any other galaxy. It lies out in the great voids between galaxies and galaxy clusters. In these voids, not only is there no one to hear you scream, but there is nothing to scream with. A cubic metre of galactic space contains approximately one million hydrogen atoms worth of matter. In a deep void, the same cubic metre contains only around one lonely hydrogen atom. To obtain matter sufficient to make one grain of sand, it would be necessary to trawl a region of deep space a thousand miles by a thousand miles by a thousand miles.

And for all this, nature is never spent. The voids are only the counterparts to massive overdensities. Gravity brings stuff together, and the gravitational force operates on a principle beloved of both hedge fund managers and the evangelist Matthew:

> 'For unto every one that hath shall be given, and he shall have abundance: but from him that hath not shall be taken away even that which he hath.' [Matthew 25:29]

More attracts more. Moving forward in time, matter has been brought together. The clumpiness of the universe has only grown with time, and it is growing steadily as we move to the future.

A more unequal future requires a more equal past. As we run the history of the universe in reverse, the allocations of matter become more proportionate. Homogeneity replaces clumpiness. First go the megaclusters of galaxies, then the smaller galaxy groups, then the galaxies themselves and finally even the stars that make them. The face of the universe becomes smoother and smoother as blemishes disappear and warts become dimples.

How do we know this? Remarkably, there is one direct snapshot of the young universe. This snapshot is called the Cosmic Microwave Background (CMB). It consists of light that was emitted when the universe was four hundred thousand years old, and which since then has travelled to us unimpeded. Having neither scattered nor interacted during its journey, it arrives to us now as a pristine image of the universe then. As the name suggests, it represents a continuous background of light at microwave frequencies, which can be measured by any suitable detector. After first mistaking their signal for the effects of bird poo, the microwave background was detected in 1967 by Arno Penzias and Robert Wilson using a distinctive horn-shaped antenna. Penzias and Wilson did not work for a university. They were instead employees of the American industrial superpower Bell Telephone Laboratories during its gilded age, when its employees won seven Nobel Prizes for a variety of fundamental discoveries.

The most striking feature of the universe revealed by the microwave background is its homogeneity. In all directions, it looks the same, at a single temperature of 2.73 degrees above absolute zero. East, north, west and south the microwave background is identical – almost. Just and barely visible, at a level of one part in a hundred thousand, are tiny irregularities. The temperatures of different parts of the sky are, in the end, not quite the same. By one hundred-thousandth of a degree, some parts are hotter and some parts are cooler. In these tiny blemishes of the past, we see our present and future. Time and gravity have grown these dimples into enormous wrinkles. What was then a small local surplus of energy is now a galactic supercluster – a colossal region of overdense space. An underdensity at the level of one part in a hundred thousand has turned into a giant void bereft of galaxies.

Such is the universe as we rewind cosmic time: younger, smaller, hotter and smoother. The more we go back in time, the more it becomes so. How far back can we go?

There is strong and more or less incontrovertible evidence that this picture holds back to the period when the universe was one second old. The temperature at this point was around thirty million degrees, and all of the universe we see today could be contained in a box around one light second – three hundred million kilometres – in size. It is at this point in the universe that the first stable nuclei were formed by nuclear reactions in the expanding universe. It is measurements of the primordial abundances of such nuclei, through observations of very old stars, that provide the support for this picture. Looking further back in time than the one second point, our equations show a universe continuing to contract. Beyond this instant, however, there is no definitive observational support for what the equations suggest.

What did happen earlier than one second? We do not know. In this epoch, all is speculation. There is good speculation and bad speculation. There are indeed ideas such as cosmic inflation, which come with a good deal of indirect empirical support. However, we do not *know*. Inflation is a plausible, indeed maybe highly plausible, idea. It may explain the detailed form of the tiny blemishes that are present in the microwave background, but it is not clear that it should be the only possibility.

There is also bad speculation. Much has been said or written about what may have happened before the Big Bang, including anthropic landscapes of different laws of physics and eternally reproducing multiverses of many possible universes. It is not that the ideas are necessarily wrong. It is rather that the extravagance of these conjectures is matched only by the paucity of either rigorous calculation or observational motivation. The danger is that this replaces physics' long-standing chaste marriage of solid theory and careful experiment with a form of scientific soft porn best suited for the pages of glossy magazines.

While we will return to this topic in chapter 10, I will for now let sobriety be the better part of speculation and be silent whereof I cannot speak.

3.7   DIFFERENT CAN BE THE SAME

I next want to describe a result that is more theoretical than experimental. It is more a discovery about the structure of physical theories than a statement about the meaning of some measured phenomenon. As such, it is not directly about the observable world in the way that the other parts of this chapter are. Nonetheless, the conceptual and intellectual significance of this discovery merits it a place in the pantheon of ideas in theoretical physics. This discovery is the existence of dualities.

In plain English a duality is the statement that one theory described one way is equivalent to another theory described a different way. Put in this form, this does not sound remarkable. There are many different but equivalent pairs where the equivalence is neither profound nor interesting. The expressions 'The fourth power of the number of green bottles originally hanging on a wall' and 'The number of men the grand old Duke of York marched up a hill' both describe the number ten thousand. However, any attempt to find a deep relationship between these two statements will end in a mental health unit. Dualities are important because they reveal profound and useful equivalences between structures and theories that on the surface appear totally unrelated.

Large numbers of dualities have been enumerated, mostly involving relations between different theories. However, some dualities – called self-dualities – can exist between the same theory with different choices for its parameters.

I describe here in some more detail how this case works and what this entails. I focus in particular on language appropriate to quantum field theories, as these are where some of the most interesting applications of duality occur. Almost all such theories have a parameter – let us call it λ – which measures the strength of interactions between different particles. This is analogous to the idea of electric charge, which measures how charged particles respond to an electromagnetic field. The trajectories of moving charged particles bend within a magnetic field, and the larger the charge, the more they bend. Particles with no charge do not interact at all and continue in a straight line.

If λ is zero, there are no interactions. We now imagine what happens as we change λ. Initially, λ is almost zero and the particles do not care about one another. They each move past with no deflection, being neither attracted nor repelled. As we slowly increase λ, the interactions become small but non-zero. Particles that pass each other deviate from their original paths. However, these paths are only slightly modified and differ merely by a small perturbation. Now increase λ further so that it becomes large, and the mutual attraction between different particles becomes enormous. The particles no longer simply pass by each other but interact strongly, with trajectories that differ totally from their original routes.

We can make a human analogy for these cases of free theories with no interactions, theories that interact only very weakly and theories that interact strongly. We replace particles with humans and consider a theory of human interactions. We imagine two people – let us call them Arabella and Bert –

walking down a street in opposite directions. Both have definite plans: Arabella is on her way to a rugby match and Bert is going clothes shopping. What happens as they pass each other? If Arabella and Bert are complete strangers to each other, they will walk by with no acknowledgments, continuing to their separate destinations. This is the analogue of the zero coupling case. If Arabella and Bert are weakly coupled – perhaps they are work colleagues – they may stop and greet each other before resuming their journeys. They may be a few minutes later than they would otherwise have been, but their destination is unchanged. The final case is the strong coupling case, where Arabella and Bert are cousins who have not seen each other for years. Here they may abandon their journeys in favour of lunch with each other and an extended catch-up. The effect of the interaction is not a small perturbation on the original state of affairs, but instead a total alteration.

It is 'easy' to predict what happens in a theory with small values of the coupling constant 'λ'. The result is a small perturbation on the case of zero coupling, and various approximation techniques can be used to evaluate this perturbation. In practice these calculations do become intricate, but the underlying principles of them are clear. However, as the coupling gets large, 'easy' becomes 'difficult', 'difficult' becomes 'hard' and 'hard' becomes 'almost impossible'. Returning to our human example, it is as if we asked what would happen in Times Square if everyone there suddenly realised they had all been at elementary school together.

Once the coupling has been ratcheted up to the 'impossible' level, what happens as we increase it further? This is where dualities enter. In a theory with a duality, something remarkable happens. Instead of continuing to ascend a gradated ladder of difficulty, we start descending. 'Impossible' becomes 'difficult', 'difficult' becomes 'hard', 'hard' becomes 'easy' and 'easy' becomes 'trivial'. The theory with infinite coupling turns out to be indistinguishable from the theory with zero coupling. The theory with coupling constant 'λ' is, in a sense that can be made precise, exactly the same as the theory with coupling constant '$\frac{1}{\lambda}$'. Such a theory is called self-dual. We can imagine turning a dial to adjust the value of one of its parameters. As we do this, we end up bringing the theory back to itself.

For this case of self-duality, the duality relates a theory in one regime to the same theory in a different regime. The other common kind of duality relates a theory in one regime to a second theory in a different regime. Here, we start with theory A at zero coupling. As you dial this coupling up, this original theory becomes progressively harder and harder to solve. However as you continue increasing the coupling, you realise that you have arrived at a second theory, B, in its own weakly coupled regime. The enormous calculational boon from this is that it allows the replacement of theory A in the 'help-I-can't-solve-it' regime with theory B in the 'small-couplings-yes-I-can-do-this' regime.

There is one example of duality that has been especially prominent in the last fifteen years. This is called the AdS/CFT correspondence, or sometimes the gauge/gravity correspondence, and was formulated by the Argentine physicist Juan Maldacena in 1997. We shall discuss this duality and its applications at greater length in chapter 6 and particularly in chapter 8. Here I shall give simply a brief introduction to what is one of the most important manifestations of duality. The AdS/CFT correspondence is an example of an equivalence between one theory in a certain regime with a different theory in another regime. It states the equivalence of a four-dimensional quantum field theory of particle interactions in the large coupling, hard-to-solve regime with five-dimensional gravity in the easily tractable weak coupling regime.[20]

The word 'equivalence' should attract arch scepticism. What is really meant here by equivalent? After all, one theory lives in five spacetime dimensions and the other theory lives in four spacetime dimensions. Furthermore, one theory involves gravitational forces and the other theory does not. Nonetheless, 'exactly equivalent' means precisely that – there exists a dictionary such that every quantity in one theory can be translated into a quantity in the other theory. Calculating a quantity in one theory gives the same result as calculating the equivalent quantity in the other theory. Information is neither lost nor gained, and there is exactly the same content on either side of the translation. As for the inscriptions on the Rosetta stone, there is the same information available however you choose to express it.

A good indication of the importance of this result is that at first sight it seems obviously wrong. The gravitational theory is in five dimensions. The gauge theory is in four dimensions. How can a five-dimensional theory be equivalent to a four-dimensional one? There is an extra dimension to the theory, and so one would expect, not unreasonably, that there are more degrees of freedom, more internal knobs you can tune, in a theory with an extra dimension.

It is true that many checks of this conjecture are not appropriate to this book, although I shall describe some at greater lengths in chapter 8. Here however, I only wish to explain, at least heuristically, why this dimensionality argument is wrong, and why it is possible for a five-dimensional gravitational theory to be equivalent to a non-gravitational four-dimensional theory.

To do so, we imagine what happens in each theory as you put lots and lots of stuff into a box of fixed size. The simplest example of a field theory is pure electromagnetism, the theory of light. In pure electromagnetism, we can in principle put as much radiation as we wish – even infinite amounts – inside a box. Such a box is called a oven. The more radiation you put in, the hotter the oven gets – but as long as the walls are strong enough, there is no actual limit on how hot the oven can be. Now consider a gravitational oven. We again take a box of fixed size and try and squeeze more and more matter into it. However, the difference is that in the gravitational theory, at some point a limit is reached. Once we reach this limit, any more matter we put in

---

[20]For afficionados of buzzwords, the gravity theory is type IIB string theory on $AdS_5 \times S_5$ and the field theory is maximally supersymmetric Yang-Mills theory at large N.

will cause the box to collapse and form a black hole. The game of feeding the oven with particles is over, and in a gravitational theory there is a maximum amount of energy that can be stuffed into a box of any given size.

Black holes are one of the best studied objects in theoretical physics, and one of their most insightful students has been the Cambridge University physicist and science icon Stephen Hawking. In 1973 he, concurrently with related proposals by Jakob Bekenstein, was able to show that black holes have an entropy given by the area of the black hole.

First, what is meant by the area of a black hole? For every black hole, there is a surface around it such that every object within that surface, no matter how fast it is moving or in what direction, will fall into the black hole. The surface is trapped: once within it, all paths lead to the singularity. The area of a black hole is simply the area of this surface, commonly called the event horizon.

Second, what is entropy? Entropy is a measure of the number of 'degrees of freedom' of an object: the total number of ways to rearrange its internals while keeping its externals unaltered. For example, the entropy of the gas in a room is a measure of the total number of way the gas molecules can arrange themselves within the room. The fact that the entropy of a black hole is proportional to its area tells us that the number of ways of rearranging the black hole internals depends upon the *surface area* of the black hole and not upon the volume. In contrast, the entropy of the non-gravitational electromagnetic oven grows, precisely as one would naively expect, with the volume of the oven and not its surface area.

This result is deep. The fact that the entropy of a black hole depends on its area tells us that in a gravitational theory the amount of stuff we can pack into a given volume is set not by the volume of the region, but instead by its surface area – which has one dimension lower. For a five-dimensional gravitational theory, as arises in the AdS/CFT correspondence, what this tells us is that the number of degrees of freedom is actually counted by a quantity which is four-dimensional and not five-dimensional.

This is very, very far from a proof of the AdS/CFT correspondence. However, it is an illustration of one way in which this apparently obviously wrong correspondence is neither obvious nor wrong.

I shall return to dualities later in this book. For now, let me restate why dualities are so important. In particular, why are they so important that, despite being entirely theoretical, they deserve a place in a list that includes quantum mechanics and relativity? Dualities matter because they have allowed the landscape of physical theories to be shrunk dramatically. Once upon a time, the strong coupling regime in the map of theoretical physics was marked as *Terra Incognita* and decorated with pictures of dragons and sea monsters. The explorers have now arrived, and they have found this land to be filled with semi-detached houses from Milton Keynes.

3.8   NATURE IS SMARTER THAN WE ARE

There is a final discovery – better, a meta-discovery – that I want to describe. This is not directly a scientific result. It is neither a physical law about nature nor a statement about the structure of physical theories. It is rather an observation founded on the other discoveries. It is an observation of the essential conservatism of nature compared to the radicalism of scientists.

It can be summarised as the statement that true revolutions are rare. Most big discoveries do not change the laws of physics, but rather reveal unexpected consequences of the existing laws. Scientific research is hard, and some of the problems tackled are very hard indeed. It is easy to spend a lot of time thinking, working and calculating, and yet not make any progress. It can be psychologically easy to give up on a problem, believing its solution requires new principles, when what in the end is required is simply clear thinking about the consequences of existing laws.

The history of quantum mechanics provides a striking example of this. The benefit of hindsight tells us that the laws of quantum mechanics as formulated in 1926 were correct. The high road that led to the Standard Model and an understanding of three of the forces of nature was marked not by modifying these laws but by understanding and applying them. However, the historical record shows that on encountering difficulties many physicists instead tried to modify the foundations. We have already discussed two examples: Einstein's troubles with quantum mechanics, and the problem of infinities in quantum electrodynamics. We can add a third to these: the problem of understanding the strong nuclear force.

While the existence of the strong nuclear force had been known since the early days of nuclear physics, its detailed experimental study only began in the 1950s and 1960s, when particle colliders obtained sufficient energy to reach the characteristic nuclear energy scales. These colliders soon discovered a menagerie of strongly interacting particles. The particles appearing in quantum electrodynamics had been limited in number: the photon (the electromagnetic force carrier), the electron and positron, and muon and anti-muon. In contrast, the study of the strong force revealed the proton and neutron, the neutral and charged pions, the neutral and charged Kaons, the K-star, the rho, the Lambda, the Omega, the phi, the eta, the eta-prime, the delta, the sigma, the xi . . . prompting the Italian Nobel Laureate Enrico Fermi to quip 'If I could remember the names of all these particles, I'd be a botanist'. The experimental manifestation of the strong force looked nothing like that of the electromagnetic force. It was entirely bizarre, and there seemed no organising logic to how and where these particles arose from. Freeman Dyson, one of the pioneers of quantum electrodynamics, had remarked in 1970 that he thought the correct theory of strong interactions would not be found for a hundred years.

The combination of unexplained experimental results and the revolutionary air of 1968 Berkeley led to further radical proposals: total nuclear democracy, the abandonment of fundamental particles and even the fusion of quantum theory and Eastern philosophy into something called the Tao of Physics. All heady, all exciting – and all wrong. In the end, the strong force is a close cousin of the electromagnetic force, and it was explained by conventional quantum field theory revealing aspects that not previously been appreciated.

It is not true that progress is never made by changing the rules of the game. The discovery of quantum mechanics is the perfect example of when this was both necessary and correct. The laws of quantum mechanics really are different from the laws of classical mechanics, and there is no way to massage the latter into the former. Quantum mechanics is, as far as we know, fundamentally right, and classical mechanics is – we do know – fundamentally wrong. The laws of classical mechanics are not the basic laws that describe this world, and there is no way to gloss this statement to make it otherwise.

This, however, is an exception and not the rule. The clash between human puzzlement and the laws of nature is generally like a clash between Chelsea and Bradford City. The result may not be guaranteed,[21] but there is only one side to bet on. Even if you are Einstein and Feynman rolled into one, nature is still smarter than you are. Almost every claim of new physics, whether the repeated identifications of the nature of dark matter or the discovery of faster-than-light neutrinos, fails. Your inability to solve a problem with the tools available is not a sign that the problem is insoluble and requires a change in the laws. Even the fact that an observation appears completely in contradiction to the known laws of physics is a very poor reason to believe that this contradiction exists.

There is perhaps a human aspect to this as well. Physicists working on understanding the fundamental laws of nature do not generally regard themselves as intellectually deficient. Excessive humility is not observed to be a common weakness within the subject. It is always hard to spend a lot of time attacking a problem, thinking hard about an idea and getting nowhere. It can be easier to believe that the inability to make progress is due to a need to reformulate physics, rather than the fault lying in a personal failure to be the smartest cookie in the biscuit tin.

A previous age would have called this pride. A modern one may assign it to the nature of psychological stimulation in early childhood. The moral is the same: there are few true revolutions. It is much easier to propose new laws than it is to understand existing ones.

With this caveat in mind, we now move on to discuss the reasons why our current theories of physics appear inadequate.

---

[21] As the 2015 fourth round FA cup scoreline Chelsea 2 Bradford City 4 shows.

CHAPTER 4

The Truth Is Out There

The previous chapter was a whirlwind tour of the great advances made in our understanding of nature. These advances are associated with the icons of physics – names such as Albert Einstein, Marie Curie, Paul Dirac or Richard Feynman. For all their achievements though, we should not be beholden to great names. However much Einstein contributed, we now know far more, and understand it far more deeply, than Einstein did even at his intellectual peak. One important aspect of this understanding is that our understanding is incomplete. The theoretical jigsaw outlined in the previous chapter has several large and crucial pieces missing. We do not know for certain what these pieces look like, or where to find them, but we do know where they fit in.

The purpose of this chapter is to explain why something conceptually new is needed. I aim to describe some of these missing jigsaw pieces – of which the first and most prominent is a quantum theory of gravity.

Some scientific theories are not expansionist. They are important in their own sphere but do not extend beyond that. The theory of fracture propagation in metals is important to anyone who boards an aeroplane, but it does not affect all of science. These Switzerlands of the scientific world do what they do well, and are content with that.

This is not the case with quantum mechanics. This is instead a jealous theory that brooks no rivals. It is not content merely with the successful explanation of known phenomena, but is instead continually seeking lebensraum. The claim that the world is described by quantum mechanics is not simply a limited statement applicable to atoms at short distances. It is instead a statement about every last bit of the world – as it was in the beginning, is now, and ever shall be. In this sense, quantum mechanics is a totalitarian theory. It refuses to coexist with alternative accounts of reality but demands for itself a universal applicability to the fundamentals of nature. In this it has been successful, and since its discovery it has spread through the population of scientific theories like a successful mutation.

The quantum field theories that so successfully describe the strong, weak and electromagnetic interactions involve a quantum mechanical treatment of the particles and the forces. The motion and interaction of the elementary particles is described with the full technology of quantum mechanics. However, these particles move against a spacetime backdrop that is taken straight from Einstein's theory. This spacetime is pristine, relativistic and classical, being entirely innocent of quantum theory.

This may have been good enough if spacetime were itself a fixed entity. Unfortunately, it is not. We saw in the previous chapter that Einstein's second great insight, general relativity, made spacetime itself dynamical. The geometry of spacetime warps and shimmers in response to the matter that passes through it. Spacetime geometry is no more intrinsically special and fixed than the electromagnetic field – which also fluctuates in response to any passing matter. Geometry then affects geometry now. Geometry now determines geometry-to-be. Small rippling perturbations in the geometry of space and time propagate outwards, just as small perturbations of the electromagnetic field – or indeed ripples on a pond – propagate outwards.

Einstein's theory of general relativity is a fully dynamical theory of gravity-as-geometry. This theory works extraordinarily well and is extraordinarily well tested. It explains the formation of black holes, the expansion of the universe and the gravitational lensing of light around massive objects. It is verified to describe the universe from an age of one second to an age of fourteen billion years. There is not a single observation that is inconsistent with it. It is a fantastic success. There is no *experimental* reason to mistrust it.

And yet, it must be wrong. General relativity is a theory untouched by quantum mechanics. It represents the last great hurrah of an older and simpler world. Once introduced, quantum mechanics cannot be restrained. The quantum theory is intolerant of its rivals; it inexorably replaces the old ideas and cannot live alongside them.

As we saw in the previous chapter, the transition from classical to quantum mechanics occurred first in the study of the motions and energies of particles, and of the measurements that could be done on them. For example, the quantum uncertainty principle forbids the simultaneous measurement of both the position and momentum of a particle. The non-relativistic quantum mechanical equations of motion were successfully extended to include special relativity. Careful study of the implications of relativity for quantum mechanics led to the description of the strong, weak and electromagnetic forces using the formalism of quantum field theory. Step by step, the laws of physics have transmogrified from classical to quantum formulations. This shift has percolated through all of physics – but not yet to gravity.

This makes it clear that something – at this stage we do not know what – is needed beyond what we already have. The known laws of physics work incredibly well. A theory of quantum gravity is required not for military applications, nor for solving outstanding industrial problems, nor by any unexplained anomaly in experimental physics. The argument that such a theory exists is purely theoretical. Classical general relativity is the last stronghold of the old physics, and all must give way before the imperium of quantum mechanics. This argument requires that the laws of gravitational physics *must* change, because the laws of gravitational physics *must* be consistent with quantum mechanics.

Whatever that something is, it must be conceptually new. In particular, we know it is not just a re-application to gravity of the techniques that have been successfully used to quantise the other three forces. A direct quantisation of gravity along this well-trodden path has been tried – and it fails. It is a noble failure, verdant with spectacularly hard and lengthy computations.

It is true that there do exist general, albeit technical, arguments why this direct approach should fail. Einstein's theory of gravity is, in the parlance of the trade, a 'non-renormalisable' theory. Maxwell's theory of electromagnetism is, in contrast, a 'renormalisable' theory. Quantum calculations in both renormalisable and non-renormalisable theories involve infinities. The difference is that in renormalisable theories, the infinities turn up in only a finite number of places. In non-renormalisable theories, the infinities turn up in an infinite number of places. In the former case, it is possible, with some work, to isolate, categorise and eliminate the infinities with the help of a finite number of experimental measurements. With an infinite number of infinities, this is no longer possible. Removing the infinities now requires an infinite number of measurements – which cannot be performed.

This argument then suggests that the head-on approach to quantum gravity should fail. This general argument does not eliminate the possibility of special structures or cancellations that evade this conclusion. This possibility can be checked only by calculation, and these calculations are hard. The problems lie not so much in intrinsic conceptual difficulties with the computation as in the enormous number of terms that need to be included.[1] This work included heroic efforts from the Texas physicist Bryce de Witt, in many ways the pioneer of this subject, and even contributions from Richard Feynman – who did not enjoy his time working on gravity, as a letter he sent home to his wife from a gravity conference in Warsaw reveals:

> I am not getting anything out of the meeting. I am learning nothing. Because there are no experiments this field is not an active one, so few of the best men are doing work in it. The result is that there are hosts of dopes here (126) and it is not good for my blood pressure: such inane things are said and seriously discussed here that I get into arguments outside the formal sessions (say, at lunch) whenever anyone asks me a question or starts to tell me about his 'work'.

This all culminated in an explicit calculation of the relevant divergences by

---

[1] One can of course reasonably view the question of how to organise an enormous number of contributing terms as a conceptual difficulty in itself.

Marc Goroff and Augusto Sagnotti in 1986. The expected divergences were there. The gravity of Albert Einstein is indeed non-renormalisable, and cannot be quantised using standard techniques.

This, more or less, establishes a need for something new. However, such negative results always require scanning for loopholes. The conclusions of any argument are only as strong as its premises, and there are always assumptions that enter a result. There is in particular one promising variation on this approach that I ought to describe, and on which the jury is still out. This variation involves not Einstein's original theory of gravity but an extended version involving many additional particles. This variation is called 'maximal supergravity'.

The majority belief is that this variation is only a sticking plaster and not a cure for the ills encountered in quantising gravity. Most probably, it does not work. If it does work, it describes a version of gravity that must be purely theoretical and cannot, even in principle, be merged with what we know of particle physics.

This variation is based on something called supersymmetry. If infinities are going to disappear, they need a good reason to disappear. In the last chapter I enthused about symmetries. Symmetries can provide a good reason for something to disappear. Symmetries make things equal. If a picture is symmetric, we know that the area of the left-hand part of the image exactly equals the area of the right-hand part, without having to work out either area precisely. A symmetry is capable of taking two halves of a calculation and forcing each part to give an answer that is exactly equal and opposite to the other part. The resulting sum of the two halves vanishes exactly, irrespective of how large each individual part may be.

It even holds if each part is infinitely large. This may sound troubling at first – the sum of positive infinity and negative infinity does not appear very well defined. It is here where a symmetry is so useful. Instead of allowing a divergence to continue all the way to infinity, we could cut off all sums at a value of ten to the power of ten to the power of one hundred thousand: a number unimaginably larger than the total number of particles in the universe. Every number in the calculation is now finite and well defined. There are no ambiguities. The symmetry still ensures that the two halves of the calculation add up to zero, and the answer has no pathological features for any value of our cutoff. As we remove this cutoff and make it even larger, the overall sum remains at zero. We can then finally take a formal limit in which the cutoff is removed altogether. As on every step of the way we have sensible answers, this last step of removing the cutoff to infinity also gives a sensible and well-defined answer.

Symmetries allow problematic infinities to be removed from calculations. This is good – as far as it goes. The problem is that generally it does not go far enough. In gravitational theories, removing infinities is somewhat analogous to playing whack-a-mole at a mole breeding farm. There are many sources of infinities. In fact, there are an infinite number of infinities to be removed. It is not enough to have a symmetry that will remove some of the infinities – it must be big enough and powerful enough to remove all of them.

One of the features that is super about supersymmetry is that it has long been recognised as being good at removing infinities from calculations. The possibility of supersymmetry as a symmetry of our four-dimensional spacetime was stumbled upon early in the 1970s, first in the Soviet Union and then independently in the West. Nature partitions particles into two classes, called *bosons* and *fermions*,[2] and at the simplest level supersymmetry relates the properties of bosons to the properties of fermions.

Simplifying slightly, in the analysis of infinities both bosons and fermions give independent contributions to these infinities. Supersymmetry is able to relate the calculational divergences caused by bosons to the calculational divergences caused by fermions, in such a way as to make them equal and opposite. The first supersymmetric theories of gravity – or supergravities – were developed at Stony Brook University in New York by Dan Freedman, Peter van Nieuwenhuizen and Sergio Ferrara in 1976, and it was rapidly realised that the presence of supersymmetry improved the situation with respect to divergences.[3] Large classes of infinities were removed – but an infinite number still remained. Gravitational theories were constructed with more and more supersymmetry; more and more divergences disappeared. However, even with the fullest and richest form of supersymmetry, the so-called maximal supergravity theory, theoretical arguments still indicated that infinities should remain, although at a level of calculational complexity that was impossible to verify explicitly. It thus appeared that large amounts of supersymmetry, by itself, ameliorated – but did not ultimately cure – the problem of divergences in the direct approach to quantum gravity. This was the situation of quantum gravity in the 1980s, contemporaneously described by Stephen Hawking's famous book *A Brief History of Time*.

Developments in the last ten years have caused this view to be revised. These developments all focus on the maximally supersymmetric theory that was considered most promising during the 1980s. One decisive result is that the old arguments as to when infinities were expected to appear are wrong. These arguments were based on the visible symmetries of the theory, and the expectation that certain infinities would remain was based on those symmetries and those symmetries alone. It has now been discovered that the maximal supergravity theory has additional, hidden symmetries that were previously missed. This extra structure ensures that many terms that had previously been believed to be infinite are actually zero – and this is enforced by the newly discovered hidden symmetries.

---

[2] As a technical aside, these correspond to particles with integer or integer-plus-one-half quanta of angular momentum.

[3] The most *profitable* development during this time at Stony Brook, however, was happening around the corner in the math department, where the department chair Jim Simons was formalising his investment hobby into the firm that grew to become Renaissance Technologies, one of the world's largest and most successful hedge funds.

This raises the question of whether *all* the infinities that may in principle arise do actually in the end vanish. Perhaps, there are further additional hidden symmetries that remain missed. Could it be the case that the visible and hidden symmetries, taken all together, are strong enough to cause the maximally supersymmetric theory to be finite, with no infinities at all? If this were the case, then there might exist a theory of quantum gravity, extending but including Einstein's theory, with no infinities at all.

The jury is still out on whether such a theory exists. The question of divergences in maximal supergravity theory must have a definite answer, but it is also not amenable to frontal assault as direct calculations are just too hard. If an answer is found, it will only be through clever arguments.

However, even if the answer is in the affirmative, it ultimately does not help so much. The prescribed medicine for the ills of infinite divergences is an enormous dose of supersymmetry. Unfortunately, this medicine is far too potent and kills more than just infinities. The same large amounts of supersymmetry that remove the infinities also make it impossible to include the Standard Model of particle physics within the theory. These whopping amounts of supersymmetry impose particular constraints on the particle types that are allowed, and the particles of the Standard Model violate these constraints.

In particular, the unavoidable consequence of such large amounts of supersymmetry is that the laws of physics must be unable to distinguish left and right. However, one of the most crucial and striking features of the Standard Model is that it violates parity – it is able to distinguish between left and right. What does this mean? As we saw in the previous chapter, this means that it is possible to define the meaning of 'left' and 'right' by reference only to the laws of physics. Suppose we were meeting for the first time distant humanoid aliens, and that we wished to start our meeting by shaking our right hands together. The laws of the Standard Model are such that we could explain this in an unambiguous way, and we could both agree on which is our right hand.

Even in principle, this is impossible within maximal supergravity. For this reason, it seems that the theory of maximal supergravity, for all its pristine mathematical beauty, is at best a technical curiosity: it is not the world, and the world is not it.

The argument of the last few pages has been a classic theoretical argument for why physics is incomplete and something new is needed. It is rather philosophical. It is abstract, founded on general principles, and appeals in at most a minor way to the actual properties of any known particles. The strength of the argument comes simply from the intrinsically quantum-mechanical nature of the world. It identifies the need for something new – a quantum theory of gravity – but offers minimal guidance as to the form it will take or the equations that will govern it. The sole requirement is that this new theory, whatever it may be, must reduce to classical Einstein gravity in the limit where quantum effects are unimportant.

This represents the ascetic aesthete's approach to realising that something more is needed. However, the same conclusion can also be reached by another, rather more empirical, route. This is based on examining the known structure of physics, and in particular the known structure of the particles, masses and interactions of the Standard Model.

These masses cover a large range. The heaviest known fundamental particle, the top quark, has a mass comparable to one atom of lead. The lightest, the neutrinos, have masses roughly one trillion times smaller. The electron is halfway inbetween: around a million times heavier than a neutrino, and around a million times lighter than the top quark.

The masses and interactions of these particles are described by the Standard Model of particle physics. The Standard Model does impose certain automatic relationships between different interactions – for example, the electron must interact with the photon in precisely the same way as a muon does. However, many of the particle masses and interactions are only parameterised but not explained by the Standard Model. They enter the Standard Model as fundamental constants – and the precise origin of these constants is someone else's puzzle.

It is indeed someone else's puzzle, because these constants have remarkable patterns and structures. Their values are not random. These constants were not drawn by gods playing dice in some pantheonic lottery. Instead, they have patterns that tell us that the 'constants of nature' appearing in the Standard Model must really be products of a deeper theory lying underneath the Standard Model.

What are these patterns? The first pattern is the existence and replication of particle families. Most of the particles of the Standard Model come in groupings called families. Each family contains a particular set of particles, and there are three such families in total. Each grouping involves similar types of particle, with the families differing only by the relative masses of the particles. The first family of the Standard Model contains the electron, the electron neutrino and the up and down quarks. The second family contains the muon, the muon neutrino, and the charm and strange quarks. The final, heaviest, family is made up of the tau, the tau neutrino and the top and bottom quarks. The particles here have been ordered by their relative type. The tau and the muon are heavier copies of the electron. The top and charm quarks are heavier copies of the up quark, and the bottom and strange quarks heavier copies of the down quark.

On hearing this, one may think that family replication could occur an infinite number of times. We have observed three families, but perhaps there is also a fourth family that is present but simply too heavy for us to observe with current accelerators.

This is wrong. It is known experimentally that no fourth family exists: the count stops at three. The simplest argument for this is that all the neutrinos are very light, and so any fourth light neutrino should already have been observed. However, this argument is not watertight – it could be the case that the fourth neutrino exists, but is different and is perhaps just very much heavier than the neutrinos of the first three families.

This loophole appears contrived but could not be eliminated for a long time. Its definitive elimination occurred with the discovery of the Higgs boson at CERN in 2012 and the analysis of its properties. Although the calculational details are technical, the existence of any fourth family, however heavy, would have drastically modified the properties of the Higgs boson away from the behaviour expected and observed in the Standard Model.[4] Fourth family behaviour was not observed, and the existence of a fourth family of particles is now decisively ruled out. There are then exactly three – and no more – particle families. This family replication, included in the Standard Model but not explained by it, is one sign of a deeper layer of physics beyond our current knowledge.

A second pointer to the incompleteness of the current framework lies in the structure of particle masses. Each jump in family, both from the first family to the second family and again from the second family to the third family, brings a jump in mass by a factor between ten and a hundred. The jump from the electron to the muon is a two-hundred-fold increase; that from the muon to the tau a twenty-fold increase. The jump from down to strange is a factor of thirty, and from strange to bottom a factor of forty. The jump from the up quark to the charm quark, and from the charm quark to the top quark, involve mass ratios of approximately one hundred and fifty. These rapidly increasing particle masses indicate a deeper underlying explanatory structure, but again it is unclear what that structure is. Something new is needed, but what is that something?

Neutrinos stand out as an exception to the above patterns. Neutrinos are special. They are almost massless and almost non-interacting. Neutrinos do, just about, interact. A neutrino passing through a sheet of lead has a finite probability of being stopped by it. This probability is not large. It requires a lottery winner's dose of good luck for one kilometre of solid lead to stop a neutrino. However, the probability is not zero. Given enough neutrinos, some will interact, and these interactions can be detected.

For example, the core of the sun produces neutrinos in abundance as a by-product of the nuclear reactions that fuel the sun. These neutrinos stream outwards in all directions. By the time they reach the earth, their flux is still such that every second one thousand billion neutrinos pass through the palm of your extended hand. Given enough target material – one hundred thousand gallons of dry cleaning fluid in the original experiment – the occasional interactions of these neutrinos can be detected.

---

[4]The underlying reason for this is that the interactions of particles with the Higgs boson are proportional to their mass, and so the heavier the fourth family is, the more strongly it couples to the Higgs boson. It is this property that implies the impossibility of decoupling any fourth family to arbitrarily large masses, as the decoupling effects of large masses are precisely cancelled by the ever-increasing strength of their coupling to the Higgs boson.

Neutrinos also do, just about, have mass. It is not possible to measure the absolute masses of neutrinos. However, it is possible to measure the difference in masses between different kinds of neutrinos.[5] These mass differences can be inferred, although in a manner that is not obvious, from the phenomenon of *neutrino oscillation*: different kinds of neutrino, such as the electron neutrino and the muon neutrino, oscillate into one another as they travel. Two such mass differences have been measured, and so we know that of the three neutrino species, at most one can be massless. The absolute mass scale of neutrinos is not known, but a combination of direct searches and limits from cosmology tell us that the heaviest neutrino can be no heavier than approximately one millionth of the mass of the electron.

The particles of the Standard Model therefore reveal a clear structure. They come in three families, and with the exception of the neutrinos, each family is heavier than the previous one by a factor lying between ten and one hundred. The neutrinos sit in almost massless isolation, more than a million times lighter than the other particles. These patterns in the particles are accommodated but not explained by the Standard Model: they are pointers to a better theory, but which theory they point to is unclear.

There is a final sign within the Standard Model of a better world to come, a sign as clear and as clean an indicator as one could hope for. One of the many parameters of the Standard Model is an angle. This angle is called the *theta angle*, and it controls certain properties of the strong nuclear force. There is one particle, the neutron, that is especially sensitive to the theta angle. As befits its name, the neutron is neutral, with no overall electric charge. However this neutrality is consistent with an internal charge distribution, where an excess of positive charge on the 'northern' side is cancelled by an excess of negative charge on the other 'southern' side.

The theta angle controls this apparent charge distribution. Its extent is determined by the value of the theta angle, as the larger the angle, the more pronounced the charge asymmetry would be. This charge distribution has been searched for – but it does not exist. To within the limits of experimental accuracy, the neutron is entirely neutral: there is no such north-south asymmetry in charge distribution. This measurement determines the value of the theta angle: its magnitude is not larger than one hundred millionth of a degree. The theta angle is, as well as can be measured, zero.

This is surprising. The theta angle is an unspecified angle, and so *a priori* can take any value between zero and three hundred and sixty degrees. If this angle were simply a randomly chosen constant of nature, with nothing to guide us we would expect its value to be somewhere in the middle of this interval. Zero at greater than a part-per-billion experimental accuracy looks suspicious.

---

[5] Pedant's corner: what the experiments actually measure is the difference in the squares of the masses.

At the level of the Standard Model, theta was just an angle, and nothing prefers one particular value over any other. It is hard to look at the measured value of this angle, though, without also suspecting hidden wheels whirring. It does not look random, it almost certainly is not random, and instead this measurement surely tells us about a lumbering hidden structure in the background revealing its shadow through the physics we already know.[6]

This second set of arguments for something new has been more practical than the first argument involving quantum gravity. It involves patterns and structures present in existing known theories. These patterns are striking, and it seems highly implausible that they occur purely by chance. Certainly the presence of the patterns implies no breakdown, either theoretical or experimental, in any existing theory, and it is logically possible that the patterns are produced randomly. However, the most economical explanation is that we, as well as our ancestors, are ignorant of some of the important truths of nature.

There is finally a direct empirical reason why something new is needed and why our existing theories are incomplete. This reason is the existence of dark matter. Dark matter is a form of matter that is implied by astrophysical and cosmological observations to make up around one part in five of the energy budget of the universe. We know a lot about what dark matter is not, but relatively little about what it is. We know it is dark – it neither emits nor absorbs light. We cannot see it, and it interacts by neither the electromagnetic force nor the strong force. Dark matter is also known not to consist of any of the particles of the Standard Model, and at best its interactions with familiar Standard Model particles are exceedingly weak.

How do we know it is even there? We weigh it. Matter matters. In the last resort, anything with any kind of energy or mass communicates via gravity, and what gravitates can be weighed. Suppose you enter a dark room with a transparent bag known to contain one kilogram of fluorescent rocks. You place the bag on scales, and the scales read two kilograms. You can see the glowing rocks, but the scales reveal the existence of an additional dark kilogram. You may not be able to see it, and you may not know what it is, but you know it is there.

The existence of dark matter is established in a similar way. The bag is at least the size of a galaxy. The fluorescent rocks are stars. The principles are the same. The mass that is weighed is bigger than the mass that can be counted, and the difference between the two reveals how much dark matter is present.

The overall mass of this galactic bag can be found in various ways. The simplest method comes from looking at the motion of stars near the edge of a galaxy. Stars orbit around galaxies in the same way that the planets orbit

---

[6]In this case, there is a prime suspect for one part of the hidden structure. The vanishing of the theta angle can be explained if the Standard Model is extended to include a very light, very weakly interacting particle called an *axion* (discussed more in chapter 10). While the axion has not been found yet, there is an active experimental program searching for its existence.

around the sun. Just as the detailed orbits of the planets tell us the mass of the sun, so the detailed orbits of stars tell us the total mass enclosed within a galaxy. This total mass can be compared with the amount of visible mass present in stars, dust and gas – and is found to be far greater, revealing the need for additional dark matter.

This technique can also be used on larger scales, applied to large clusters of hundreds or thousands of galaxies. It is now the orbits of galaxies rather than stars that are useful – and these again tell us that the total mass pulling on the galaxies is far larger than the visible mass present in either the galaxies, interstellar gas or dust.

An entirely complementary way of measuring the total mass present is through examining the bending of light. Einstein's theory of general relativity tells us that heavy objects attract light, and that the paths of light rays are bent when they pass massive objects. The more massive the object, the more the direction of the light is changing. Through studying the distortions in light paths caused by the gravitational bending and lensing of light, the total mass of an object can be inferred.

The study of gravitational lensing has led to one of the most striking pieces of evidence for dark matter, in the so-called Bullet Cluster. The Bullet Cluster is the aftermath of a collision between two large clusters of galaxies. In this collision, the familiar matter within each cluster has collided, leading to the formation of a visible shock wave where the two clusters have passed through each other. The lensing techniques can however also map the total mass distribution – including the dark matter. The dominant dark components have passed through each other smoothly without colliding, and are well separated from the shock wave of ordinary matter. There is no evidence for any interactions between the dark matter components – but they can still be traced through their ability to bend light.

In this way, dark matter is something defined largely by an absence: it is the difference between what is weighed and what is seen. It is natural to wonder whether this really counts as good evidence for the actual existence of dark matter. For example, some 'failed stars' such as brown dwarfs can be compact, massive objects that emit almost no light. However, these do not require any new particles or new physics: they are just known astrophysical objects, made up of familiar constituents, that happen to be both massive and dark. Another possibility might be small black holes: black holes are also dark objects that do not require any new forms of matter for their creation.

While once weighty, these objections are no longer sustainable. The existence of dark matter is no longer inferred simply from the rotational properties of stars in galaxies. Even in the earliest epoch of the universe, when the universe was only a few hundred thousand years old, the existence of dark matter can be established through the spectrum of light present in the cosmic microwave background. This spectrum contains within it evidence for acoustic sound waves in the early universe, oscillations that are analogous to the vibrations of a drumhead and are imprinted within the structure of the microwave background. These oscillations arose in the early universe in a bath of protons, electrons and light. The details of this spectrum tell us how much matter was present in the form of protons and electrons – and how much was not. The part that was not represents new matter with at best very weak interactions with familiar particles – dark matter. In this way, the cosmic microwave background provides additional, independent, evidence for the existence of dark matter.

What do we learn from the three jigsaw pieces in this chapter? We learn that the framework of modern physics, the Standard Model plus general relativity, is not and cannot be complete. It is not complete theoretically: it is a hybrid between quantum and classical theories of forces. It is not complete internally. It contains patterns and structures that must arise from a deeper underlying theory. It is not complete experimentally: it provides no suitable candidate for the dark matter that pervades the universe. We realise that what has been achieved is magnificent, but also that magnificent is not enough – something new is needed. We do not know what that New Thing should be, or what form it should take, but what we know does tell us that it must exist.

None of this serves, as such, as an argument for string theory. It is instead an argument that we cannot rest on our laurels. Physics is not finished. New structures must exist, and it is important to explore what these new structures might be. If string theory is anything, it is a consistent theoretical structure that connects to known ideas in physics while also extending them. We know that deeper truths about nature are required. While not guaranteed, many think string theory will represent some part of these deeper truths.

It is now time to start describing string theory.

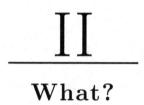

II

What?

CHAPTER 5

What Was String Theory?

5.1 THE BIRTH

What is meant by string theory?

It is 1968. In the United Kingdom, England are World Cup holders and the last mainline steam trains take their final journey from London to Carlisle. Enoch Powell speaks about immigration, evoking the river Tiber foaming with much blood, and is dismissed from the Conservative front bench. In Paris, students riot, and in the Vatican Pope Paul VI issues the encyclical Humanae Vitae. In the United States, the Standard and Poor index closes above 100 for the first time, while the rich scents of marijuana and free love merge with chants of 'Ho Ho Ho Chi Minh, the Viet Cong are going to win!'

At CERN in Switzerland, a young 26-year-old Italian physicist named Gabriele Veneziano is trying to understand the properties of the strong nuclear force. The strong nuclear force is at this time entirely baffling. A large number of strongly interacting particles have been discovered, but there is a lack of an organising principle for how they behave. Veneziano is trying to understand how strongly interacting particles scatter off each other. In particular, he is thinking about what happens when two pion particles are collided to make one pion and one omega particle. He realises that a single formula can capture many of the features of this process – a single formula that involves a famous function from mathematics, the beta function of Leonhard Euler, the greatest mathematician of the 18th century. Veneziano writes a paper, published on the 1st of September 1968 in the Italian scientific journal *Il Nuovo Cimento*, in which he puts down this formula, suggests it may describe the strong interactions, and studies many of the interesting properties possessed by it.

The article was about the messy properties of the strong force. The words 'string' or 'strings' appeared nowhere within it, and quantum gravity was not even within the outskirts of the penumbra of the concepts discussed. Nonetheless, this paper was the first-ever paper on string theory, and it acts as an intellectual Eve from which all subsequent work on the subject is descended from.

The formula Veneziano wrote down during that heady summer of 1968 is now called the *Veneziano amplitude* – the word 'amplitude' being the conventional scientific term to describe how likely it is that particles will scatter off each other. Even at the time, Veneziano knew he had discovered something important – and so did others. One of the early pioneers of string theory, Joel Shapiro, described it as follows.

> This paper arrived at the Lawrence Radiation Lab in Berkeley in the summer of 1968 while I was away... and I returned to find the place in a whirlwind of interest. Everyone had stopped what they were doing, and were asking if this idea could be extended to a more accessible interaction, such as $\pi\pi \to \pi\pi$ [the scattering of two pion particles from each other].

Gabriele Veneziano had found a key, and at the time he and others thought that the lock this key turned was in the door to understanding the strong interactions. There was an immediate rush to generalise his formula to other processes, such as the scattering of two pions into two pions, or to extend it further to the scattering of two pions into three or four pions.

The Veneziano amplitude described the scattering of two particles into two particles. Whatever theory the amplitude was part of, this theory should also describe the scattering of two particles into three particle, or three particles into three particles, or twelve particles into seventeen particles. One target was there: generalise the Veneziano amplitude into an N-particle amplitude. Within months, this problem had been solved by several independent groups, to give formulae that would describe the scattering of N particles within whatever theory Veneziano's amplitude described.

These results were part of an explosion of interest in this topic, as it became clear that a new theoretical structure had been identified. In various places groups of physicists turned their thoughts to these ideas – in particular at CERN, where a large group under the inspirational and charismatic leadership of Daniele Amati attracted interested minds from around the globe. At first this structure was called the Dual Resonance models. These Dual Resonance models captured some – not all – aspects of the strong interactions, and the models appeared to have some internal problems, but it was hoped that further work would see a solution to these issues.

The Dual Resonance models defined a relativistic quantum theory and contained lots of structure that required more study. It is important to remember that these Dual Resonance models were still exactly that: 'string theory' did not exist yet, and the only professed motivation for studying these models was for their ability to describe the strong interactions.

It was next necessary to understand the conditions required for the consistency of this structure. It was well understood that quantum theories contain consistency requirements that do not apply to classical theories, but the procedure was thought to be straightforward: there should be no 'unphysical' states in the spectrum and no negative probabilities. Through 1970 and 1971 many physicists worked on understanding these new structures, and one of them, Claud Lovelace, came up with a curious and unexpected result: the Dual Resonance models were indeed consistent quantum theories – precisely if they were formulated in twenty-six dimensions (twenty-five spatial dimensions and one time dimension). With hindsight, this was the first inkling that the theory under consideration would have to involve extra dimensions.

Towards the end of 1971 Lovelace accepted a faculty position at Rutgers University in New Jersey. His result about twenty-six dimensions did not lead to him being regarded by his colleagues with awe. In Lovelace's words,

> I was the only professor not being promoted despite the many citations of my papers. However, the jeers of the physics establishment did have one good consequence. When my discovery of the critical dimension turned out to be correct and significant, they remembered that I had said it first. One has to be very brave to suggest that spacetime has 26 dimensions.

We would now say that the discovery of the Veneziano amplitude initiated the first great wave of study of 'string theory', from 1968 to 1973, marked by a study of the Veneziano amplitude, its extensions, and their properties. All of the above is now viewed as the first epoch of string theory. However, for much of this period, the response to the question 'What is string theory?' would have been 'String theory? What have strings got to do with anything? We are studying amplitudes to describe the strong interactions!'

5.2   THE NAMING CEREMONY

In science, everything is clear in retrospect. Once you know what is essential and what is a distraction, the path to illumination is easy. Without these benefits, periods of muddle are necessary preliminaries for important results. Discoveries are only major if they resolve previous confusion, and the best researchers seek to maximise, not minimise, their number of 'How could I have been so bloody stupid!' moments.

It is therefore strictly the advantage of hindsight that makes it clear that the Dual Resonance models really were a set of amplitudes for the scattering of quantum relativistic strings from each other. However this realisation did not come until some time after interest in them was well established. It arose independently through the work of Yoichiro Nambu, Holger Bech Nielsen and Leonard Susskind, around 1970 and 1971. These three men came from very different backgrounds. Nambu was born in Japan in 1921 and had to learn his physics among the chaos and destruction of first wartime and then occupied Tokyo. He spent three years having to sleep on a straw mattress on his office desk. Nielsen was Danish, born in 1941 soon after the Nazi occupation of Denmark. Susskind, born a year earlier in 1940, was a product of the straight-talking Bronx Jewish community in New York. The scientific careers of these three men would later go in interesting directions. Almost forty years later, Nambu was the winner of the 2008 Nobel Prize for physics; Nielsen was exploring ideas where a time travel conspiracy would cause the universe to prevent the discovery of the Higgs boson; and Susskind was writing a popular book, *The Cosmic Landscape*, explaining how the anthropic multiverse was the scientific answer to intelligent design.

The realisation that there was an underlying theory of strings was not instant and came in stages. How did it occur? Analysis of the Veneziano amplitude made it apparent that the theory it described included particles of progressively increasing mass. From it, it was possible to read off both the number of particles involved and their masses. The first clue to the string-like nature of this amplitude was the fact that the particle energies (given Einstein's identification of energy and mass) inferred from the Veneziano amplitude were also the energies that arose from studying the vibrational energies associated to a string.

Where do the energies of a string come from? A string can be plucked, and can oscillate, in every direction transverse to its length. A string living in two spatial dimensions has one direction it can oscillate in; a string in three spatial dimensions has two directions it can oscillate in; a string in twenty-five spatial dimensions has twenty-four directions it can oscillate in. A plucked string can oscillate both at its lowest, fundamental frequency and also at higher frequencies – harmonics, which can be excited along any of the plucked directions. There is an energy associated to each harmonic: the higher the frequency, the greater the energy. The first inklings of strings came from the fact that counting these harmonics and their energies revealed a precise match with the particle content and energies of the Veneziano model.

'String theory' slowly began to emerge as in 1971 first Nambu and then Tetsuo Goto wrote down a set of equations from which the Veneziano amplitude would emerge. These equations were those of a relativistic string, and were the starting point from which a systematic calculation procedure could be developed. The Veneziano amplitude had been found essentially by guessing an answer. The Nambu-Goto equations for a string now provided a set of principles from which the Veneziano amplitude could be derived.

We have already seen in chapter 3 that quantum mechanics is more constraining than classical mechanics: a consistent classical theory does not necessarily imply a consistent quantum theory. Was the Nambu-Goto string another such case? It was a nice classical theory, but did it make sense in quantum mechanics? The quantum consistency of the Nambu-Goto string was established in 1973 by Peter Goddard and Charles Thorn.[1] Their result

---

[1] Following a conventional academic career, Peter Goddard became Master of St John's

was called the No-Ghost Theorem, and what this exercise in ghostbusting involved was showing that the quantum string was indeed consistent and had no physically meaningless configurations – which are known as ghosts for obscure technical reasons. This result held precisely when the number of dimensions was twenty-six, thus confirming the older result of Lovelace for the Dual Resonance models.[2]

The development of string theory also allowed a further loose end from the Dual Resonance models to be tied up. In addition to the Veneziano amplitude, another amplitude, similar in spirit but different in detail, had been discovered by Joel Shapiro and Miguel Virasoro. This amplitude also described the scattering of two particles into two particles. It shared many similar properties with the Veneziano amplitude, and it also required twenty-six dimensions for consistency. However, this amplitude was different in detail, and it was clearly not describing the same process.

It is an elementary observation that strings come in two kinds – those that join up back to themselves, as in a loop, and those that have endpoints, such as violin strings or shoelaces. The former are called closed strings and the latter are called open strings. With study, it became apparent that the Shapiro-Virasoro amplitude described the behaviour of closed strings while the Veneziano amplitude was the one appropriate for open strings that had endpoints: as the strings involved were different, the amplitudes took slightly different forms. In this way both the Veneziano amplitude and the Shapiro-Virasoro amplitude could be understood as arising from theories of strings – and with this realisation, string theory was born and came into its patrimony, while 'Dual Resonance models' had to subsume themselves into this broader framework.

If asked what 'string theory' was at this stage, however, the form of the answer would have been the same as for the Dual Resonance models: it was a candidate theory for understanding the strong interactions. This candidate theory was about to be flattened though by the combination of new experimental data and a new theory of the strong force, quantum chromodynamics.

It was not just a new social and cultural outlook that was brewing throughout the late 1960s and early 1970s. New accelerators were also being constructed across the world. In California the Stanford Linear Accelerator, operating from 1967 onwards, was accelerating electrons to higher energies that had ever been achieved before, and then colliding them with protons and neutrons. By doing so, it was probing the inner structure of the proton and neutron, the particles held together by the strong force. At CERN the Intersecting Storage Rings had started operating in 1971. These were at the time

---

College in Cambridge, before moving to become director of the Institute in Advanced Study in Princeton. He lectured me during my undergraduate studies in Cambridge, and invited the class for a party to the Master's Lodge – where I learned of his impressive collection of model hippopotami. Charles Thorn is now a senior professor at the University of Florida.

[2]The No-Ghost Theorem will reappear again in chapter 9 as part of the discussion of the monstrous moonshine program in mathematics.

a technological marvel. Instead of accelerating one beam of particles and firing it at a stationary target, it was able to circulate two distinct beams, one clockwise and one anticlockwise, and focus and collide them into each other. These accelerators were able to probe the strong force at higher energies than ever before, and in doing so, led to the discovery of a novel phenomenon – hard scattering.

Colliding particles always have an axis along which the particles approach each other. Hard scattering occurs when particles emerge both at large energies and large angles from the collision axis. Its opposite is soft scattering, where the particles continue mostly on their original paths and are only deflected by small amounts. Hard scattering is also the distinctive feature of point-like scattering. If a BB gun is fired at jelly, the shot will pass through in the general direction it was fired. However if the same gun is fired towards a set of marbles, the shot may ricochet at large angles.

In a previous age, it had been the existence of hard scattering that had enabled Ernest Rutherford to discover in 1911 that the atomic charge was concentrated within a nucleus, rather than being diffusely smeared throughout the atom in the plum pudding model. Rutherford had fired energetic particles at a thin sheet of metal and had seen then bouncing back. It was this (surprising) observation of hard scattering that forced Rutherford to the conclusion that the atom contained an almost point-like nucleus:

> It was quite the most incredible event that has ever happened to
> me in my life. It was almost as incredible as if you fired a 15-inch
> shell at a piece of tissue paper and it came back and hit you.

Rutherford's story was now being relived almost sixty years later. It was the surprising observation of hard scattering in strong interaction events that led the physicists of the 1960s and 1970s to the conclusion that the atomic nucleus itself was also made up of pointlike components – *partons* or *quarks* – and point-like objects are not string-like objects.[3]

Morally, the string model of the strong force was like the plum-pudding model of the atom – a model in which 'charge' was smeared out across an extended region. This implied a distinctive property of the Veneziano formula and its relatives – as you increased the energy of collisions, scattering would become softer, and in the limit of extremely high energy, hard scattering events would become extremely unlikely. As throughout 1972 and 1973 the collision energies were increased, the data however told the reverse story. Furthermore, the data appeared precisely consistent with the new theory of quantum chromodynamics – and as more data came in, the characteristic predictions of quantum chromodynamics became better and better fits to this data. Even before the results became decisive, the scent of the right theory was picked up. Leonard Susskind recalls,

---

[3]The language used was a war of the words between Richard Feynman and Murray Gell-Mann. Feynman called them 'partons', and pretended not to understand the word 'quark'; Gell-Mann took the reverse attitude.

At a physics conference [in 1974] I asked, 'You people, I want to know your belief about the probability that QCD is the right theory of hadrons.' I took a poll. Nobody gave it more than 5 percent. Then I asked, 'What are you working on?' QCD, QCD, QCD.

The data had surged in one direction, while the string theory of strong interactions – Dual Resonance models – was pointed along the opposite axis. String theory as an account of the strong interactions received a final *coup de grâce*: the new theory of quantum chromodynamics could also explain why string-like models had succeeded in giving an approximate description of strong interactions. In quantum chromodynamics, the force lines of the strong interaction did indeed tend to bunch into strings, in a similar way that the magnetic field lines of electromagnetism bunch together. The hadrons and mesons of the strong interaction, which the Veneziano model had been created to describe, *did* have an approximate description as points of strong charge bound together by tubes of strong force flux – and so resembling relativistic strings under tension. This explained why the string theory of strong interactions had captured certain features of the data but also could never be more than an approximation to the true theory.

As measured by the number of papers published, interest in string theory died as rapidly as it had risen. As in 1974 Abba won the Eurovision song contest with *Waterloo*, Johan Cruyff and the Netherlands dazzled the world with Total Football, and Gerald Ford replaced Richard Nixon as president, a question to most physicists of what string theory was, and what it was for, would have received a bleak answer: string theory was a failed theory of the strong interactions, and it was good for nothing.

5.3 THE WILDERNESS YEARS

However, for string theory this was not the end – nor even the beginning of the end. With hindsight, it was only the end of the beginning. In the period from 1968 to 1973, the attraction of the Veneziano amplitude, the Dual Resonance models, and subsequently string theory was not simply the chance that they might provide a description of the strong interactions. Another attraction was the realisation was that there was an underlying structure that was at once both interesting and non-trivial. There was clearly *something* there. This happens sufficiently infrequently in physics that, when it does, people pay attention. Most of physics is about modifying or exploiting known theories in known ways, in the hope that the modification can explain data that is not otherwise explained. The appearance of genuinely new structures or new theories is not so common.

It is worth digressing here about what is meant by a theory or a structure. The terminology is loose, and I do not intend to be doctrinaire, but I do want to avoid some misunderstandings. What is a theory not? It is not just a guess, whether simple or educated, as in 'my theory for why the bucket is in the bedroom and not in the kitchen is that one of the children was using it as a policeman's hat'. Neither is it only a possible explanation for an experimental anomaly, to be resolved by more data – 'model' is the term used here. It is more a chain of ideas and equations that is glued together by some combination of physical and mathematical reasoning. By itself, neither mathematics nor logic is enough. A purely logical chain tends to be far too brittle – if a single step fails, then the entire argument collapses. A good physical theory is more robust, being able to survive errors of thought and interpretation to still give useful results.

It may seem a bit paradoxical to say that logic is insufficient. After all, is not logic the principal component of clear thinking and good argument? Actually, beyond a certain elementary level the answer is no. The rules of logic are to good science what the rules of prose as taught at primary school are to good writing. Why is this so? As an example, consider the sum

$$\sqrt{\frac{6}{1^2} + \frac{6}{2^2} + \frac{6}{3^2} + \frac{6}{4^2} + \dots}$$

If you evaluate this sum for the first one hundred thousand terms, you will find that it equals 3.14158. We also remember that π starts its expansion as 3.14159. Now, there is no *logical* reason why a series whose first hundred thousand terms sum to a number looking suspiciously like π should have anything to do with π. The fact that the expansion appears to be tending to π – there is no *logical* reason why this should be anything other than an accident.

Now, in fact this series does converge exactly to π, and this can be shown with the full panoply of formal pure mathematics. However, one should not require this to give serious credence to the idea that the series does indeed sum to π. The numerical coincidence should be sufficient motivation to devote time to investigating further the properties of the sum. While this was a deliberately simple example, the general (true) lesson is that patterns should attract attention, even if they have no strict logical significance.

The everyday life of theoretical physics is full of analogous examples, and it is not formal logic that is used to buttress the conclusions. Instead, it is supporting evidence, the use of multiple different lines of argument to obtain the same result, the presence of unexpected cancellations and simplifications, and the occurrence of Goldilocks calculations where everything turns out 'just right'.

One of the principal attractions of first Dual Resonance models and then string theory was the feeling that a Goldilocks structure had been encountered. Although the direction it led was unclear, the consistency of the theory was 'just right'. Calculations worked – but only just, imposing consistency requirements that could be satisfied with nothing to spare. It was this sense that kept people working on string theory even when it was clear it had failed as a theory of the strong interactions. John Schwarz describes this period:

[We] felt that string theory was too beautiful to be just a mathematical curiosity. It ought to have some physical relevance. We had frequently been struck by the fact that string theories exhibit unanticipated miraculous properties. What this means is that they have a very deep mathematical structure that is not fully understood. By digging deeper one could reasonably expect to find more surprises and then learn new lessons.

While the majority answer in 1974 to 'What is string theory?' would then have been 'A failed theory of the strong interactions', to some the answer was 'An intriguing framework that still needs exploring'. To many of those who had worked on the subject from 1968 to 1973, something had been found – and the question was what that something was. The answers in 1974 were fragmentary. The 'something' was certainly a theory of strings. The Dual Resonance models had been successfully reinterpreted as arising from a theory of quantised strings. The scattering amplitudes that arose from the Dual Resonance models were those of the scattering of strings. The 'something' also appeared to involve more dimensions than the ordinary four, as these original string theories, called *bosonic string theories*, required a total of twenty-six spacetime dimensions for consistency.

However, many puzzles still existed. Not the least of these puzzles was the fact that string theories always contained an unphysical particle, called a tachyon. The equations for a tachyon naively make it appear that it is a particle travelling faster than the speed of light. This is not however the right way to think about them. Really, a tachyon signals an instability in a theory, like the case of a ball balanced precariously at the top of the mountain. Given the slightest perturbation, the ball will roll down and away. The only thing unphysical about this ball is the assumption that its position is a permanent state of affairs. Its location is perfectly physical, just unstable – and the ball will not long survive there.

The presence of the tachyon indicated that the bosonic string was also unstable. Despite all efforts, no direct solution to this problem could be found, until in the early 1970s first Pierre Ramond, and then André Neveu and John Schwarz, made a modification to the bosonic string equations to create the *superstring*. The essence of the modification was to include additional fermionic particles into the theory. The modification also crucially required an additional symmetry – *supersymmetry*, which has subsequently played an important role in models of particle physics.

Compared to the bosonic string, the equations and properties of the superstring were similar but better. In the superstring, the tachyonic particle and its instability were absent – although this was not something that was entirely obvious at first. Furthermore, technical reasons implied that the critical number of spacetime dimensions also changed – the Goldilocks number was now ten rather than twenty-six. The superstring also contained in its spectrum a massless particle with two units of intrinsic quantum angular momentum. These were known to be the properties of the *graviton*, the hypothesised particle that plays the same role for gravity as the photon plays for electromagnetism. In 1973 and 1974 Tamiaki Yoneya, and separately Joel Scherk and John Schwarz, showed that the interactions of this stringy particle were indeed the same as the graviton of general relativity.

With hindsight, these properties were pointing towards the reinterpretation of string theory as a candidate theory of quantum gravity rather than a candidate theory of the strong interactions. However, this change of perspective did not come easily. As happens frequently in science, the change was preceded by a period of first trying to force the theory into the answers that were desired, rather than looking at what the theory actually said. For example, the change in perspective also involved an enormous change in energies. Strings have a tension. Tension is measured in energy per unit length. The shift in target from strong interactions to gravitational interactions involved a shift in the magnitude of this tension by a mere factor of one hundred billion billion billion billion, or 10^{38}.

The proposal that string theory should be viewed as a fundamental theory of gravity was first put in print by Scherk and Schwarz in 1974. While this represents the modern view, this proposal was not received as if it had come down from Mount Sinai. The development of a quantum theory of the gravitational force is the historical *problema di tutti problemi* of theoretical physics: string theory had already failed once at a simpler task, and there was no reason to expect success simply through doubling down and multiplying the stakes.

In any case, few had motivation to care. The 1970s was also the decade of the triumph of the Standard Model. Its basic structure was confirmed and reconfirmed as new particles were discovered – the charm quark in 1974, the tau lepton in 1975, the bottom quark in 1977, and the gluon, the force carrier of the strong interaction, in 1978. The predictions of the Standard Model were verified in experiment after experiment. As energies increased, jets of strongly interacting particles – a key prediction of quantum chromodynamics – were observed in the aftermath of particle collisions. Such was the confidence in the success of the Standard Model that at the end of the decade, some of its developers – Sheldon Glashow, Abdus Salam and Steven Weinberg – were awarded the 1979 Nobel Prize, even though it would still be another four years before the force carriers of the weak interactions, the W and Z bosons, were discovered in 1983.

For theorists who had stuck with quantum field theory through the 1960s, this was a period when everything they touched turned to gold. Emboldened by their prior success, their thoughts started turning upwards. Was there a unification principle behind the Standard Model? At large enough energies, could the forces of the Standard Model simply be different aspects of the same force? If this were true, all the interactions of the Standard Model would be merely different facets of a single theory. Many (grand) unification models were proposed using quantum field theory, starting with a 1974 paper by Howard

Georgi and Sheldon Glashow of Harvard.[4] These theories predicted that the proton, the particle at the centre of the hydrogen atom and therefore necessary for life as we know it, was unstable, decaying with a lifetime vastly longer than the age of the universe. Ambitious experiments were proposed to look for these predicted decays, with the expectation of imminent success. For those who worked on understanding and predicting experimental results, this was a time of triumphalism.

Amidst an atmosphere of total indifference a few people continued research on string theory throughout the 1970s.[5] It was not a fashionable topic and string theory was not an easy subject to make a career in. However, those who remained had a free choice of problems. In 1988 the British physicist Michael Green – now recently retired from Isaac Newton's old professorship, the Lucasian Chair of Mathematics at Cambridge – recalled this era as follows:

> In a sense life was very nice in those days because particle physics is generally a very competitive subject and it was just nice to be working on something that we could take at our own pace without feeling pressurised.

String theory in the later 1970s was not about experimental data. Instead, it was about as far removed from data as it was possible to be while still remaining within the umbrella of science. It was not even really a theory. It was more an embryonic hope budding inside a chrysalis. It was a hope that this poorly understood set of ideas, after resolving the many uncertainties that still accompanied them, might just possibly maybe give a quantum account of the gravitational force.

However, even if this best-case scenario were true, it seemed entirely implausible that there would ever be any way to test it, given the enormous energy scales involved in quantum gravity. In short, string theory in the 1970s was theory for theorists. The problem it was concerned with had almost no empirical consequences. The solution, if found, appeared entirely untestable.

There is nothing wrong with this. Scientists are different, and different scientists are good at different things. Some problems require years of closeted monk-like attention, while other problems require an obsessive focus on the latest data. Some problems can be elegantly solved with pen and paper, whereas others can only be tackled by techie code wizards harnessing hundreds of computers. Some scientists want to dominate others, intellectually and socially, while others prefer to walk their own path in their own way. The belief in a single approach to good science results in only a single style of science getting done.

---

[4]To quote the Stanford physicist Michael Peskin on this work: 'The remarkable hubris of this paper makes it required reading for every student.'

[5]Ten years of apparent deadend deadbeat research on an area abandoned by the bright young things: it is fortunate that the university administrations of the 1970s were focused more on being universities and less on the outcome-centred impact attainable through leveraging transformative interdisciplinary synergies with stakeholder partners.

So string theory research in the late 1970s and early 1980s – in that it existed at all – consisted of untangling and understanding the structure of the theory. It was also clearly a topic on the boundaries of what was respectable. In the meantime, several new theoretical ideas appeared and commanded much attention. Although not obvious at the time, it would subsequently turn out that much of this apparent mainstream would end up as tributaries that fed into string theory.

5.4   HARBINGERS OF GREAT JOY

One mainstream idea of the 1970s was that of supersymmetry, which over the course of the decade went from nothing to become a large part of theoretical physics. We have already seen in chapter 3 the importance attached to symmetries in physics. There are two basic kinds of particles, bosons and fermions. The distinguishing feature of supersymmetry is that it transforms bosons into fermions and fermions into bosons. Under this symmetry, the interactions and masses of bosonic particles are related to those of fermionic particles. In a sense that can be made precise, this symmetry relating bosonic and fermionic particles can also be viewed as an extension of the relativistic symmetries of space and time – and in this same sense, it is also true that supersymmetry is not just *an* extension but *the* extension. Supersymmetry is the only possible allowed extension of the spacetime symmetries of relativity.

Supersymmetry first appeared in the early 1970s – in fact, one of its first appearances was within string theory itself. Once the idea of supersymmetry appeared, it began to percolate, being applied to any possible available theory. First, it was applied to particles. After particles, the next step was field theories – a process carried out from the mid-1970s. There was a machine and recipe to follow, and supersymmetric theories were constructed wherever they could be. After field theories the next step was gravity, and supersymmetry was soon being applied to general relativity, producing what are called *supergravity* theories. The German physicist Werner Nahm classified all possible supergravity theories in 1977. In particular, he showed that the maximal number of dimensions that supergravity theories could exist in was eleven – leading to the explicit construction of the biggest and baddest supergravity theory of them all: the writing down in 1978 of eleven-dimensional supergravity by Eugène Cremmer, Bernard Julia and (again) Joel Scherk.[6] This was the apex: Nahm's result implied that it was not possible to go beyond the eleven-dimensional theory.

Unification of the gauge forces into aspects of a single force was also one feature of the 1970s. However, the rise of the supergravity theories – which appeared increasingly special and unique as the number of dimensions

---

[6]Those whom the gods love die young – the story of Joel Scherk is, like the story of another great Frenchman Évariste Galois (1811 – 1832), a tale of so much done and with so much still to do. Scherk died tragically young in 1980 at the age of 33 – although not in a duel like Galois!

reached double figure – suggested another possible type of unification: unification through geometry. The dream that geometry is responsible for the laws of physics has been passed down from Einstein. The supergravity theories offered a possible way in: the geometry of the extra dimensions would determine the physics of the lower dimensions.

The ability of extra dimensions to turn gravitational interactions into gauge interactions had been known for a long time, going back to the work of the German physicist Theodor Kaluza. Kaluza had originally showed, as far back as 1919, that Einstein's gravitational theory in five dimensions, looked at from a four-dimensional perspective, behaved as a four-dimensional theory of gravity with an additional gauge force resembling electromagnetism. This idea had long lay dormant, but it could now find a natural home in the higher dimensional supergravity theories that were being developed. Edward Witten, recently tenured in Princeton and still on the early parts of a trajectory that would see his name become one of the most revered in the subject, caused some excitement in 1981 by showing that eleven was not only the largest number of dimensions one could obtain in supergravity, but also the smallest number of dimensions that were required to obtain the forces of the Standard Model using the approach that Kaluza had pioneered.

Over this period, research in string theory proceeded slowly. While the supersymmetric string had been developed by Pierre Ramond, André Neveu and John Schwarz, it was still not yet clear that this was a theory that fully made sense. In 1976 a modification of this was made by Ferdinando Gliozzi, Joel Scherk and David Olive, and in fact this modification produced the first fully consistent supersymmetric string theory. They also realised that the string theory contained quantum field theory, as a limit in the case where all the energies were much smaller than the characteristic energies of the strings. This was the indication that string theory contained both supersymmetry and quantum field theory within it, thereby linking the subject to the developing mainstream trends. However, the significance of this paper was something that was only appreciated later. At the time its ripples were extremely small, fading out quickly.

The paper of Gliozzi, Scherk and Olive had considered the cases of both closed strings and also open strings with endpoints. Although no one noticed then, the calculations were only correct for the case of open strings. It was not until 1981 that Michael Green and John Schwarz rectified the mistakes for the case of closed strings. Green and Schwarz also discovered that there were two possible consistent string theories in ten dimensions, which they called type IIA and type IIB.

It is actually a bit of an exaggeration to say that they knew at this time the theories were consistent. It was known that there were many subtle issues, and a full understanding of the deeper structure was still rather foggy. These ten-dimensional string theories presumably had a connection to the ten-dimensional theories of supergravity – but only one of these supergravity theories (corresponding to the type IIA theory) had actually been constructed at the time. Following Nahm's 1977 classification, it was known that the other supergravity theory – type IIB – should exist in principle, but no one had yet written it down explicitly or determined the precise form of its equations. Indeed this was not fully accomplished until later work by Paul Howe and Peter West in 1984.

The above material has been dense, so let me step back and review where it has led to. Research in string theory in the later 1970s and early 1980s, in that it existed, was focused on understanding the theory of quantised, relativistic strings. It was not focused on applications and it was certainly not focused on matching data. It was about understanding what the theory was and what consistency conditions applied to it. This required learning the relevant calculational tools and how they could be used. There was an implicit hope that string theory could represent a theory of quantum gravity, but this hope was backed by little hard evidence. During the later part of this period, string theory began to connect back into the more mainstream parts of physics, and in particular connections appeared with supergravity theories, which were then regarded as the most promising candidates for unified quantum gravitational theories. At this stage string theory was fifteen years old, but still a very minor part of theoretical physics. While a few people were interested, it was – like popular music and the stage – not an area where a young person could be advised to make a career in.

String theory moved from the wings to centre stage in 1984, starting a period that is often called the first superstring revolution (although this name could with equal justice be applied to the period from 1968 to 1973). This year has become part of folklore for what is ultimately a sociological rather than a scientific reason – the sheer rush of people that started working on string theory then, as well as the sudden transition of the subject from a poorly studied curiosity to the most active and topical subject in theoretical particle physics. The scientific content would have been identical had the transition been slower, with a more gentle influx of numbers, or had there been greater prior interest in string theory such that its change in status was less marked. However all of these would have reduced the human impact of the story: no billionaire is quite so interesting as one who was penniless only a few months earlier.

It is worth unpacking the reasons for this change in attitude. At the time, supergravity theories had been regarded as the best approach towards both quantum gravity and possible unifications of gravity with the other forces. The volte-face came about through the realisation that these supergravity theories had severe problems – and string theory had just the right structure to solve them. These problems were of two sorts. The first was a problem of finiteness, and the second was a problem of consistency. The problem of finiteness was that when supergravity was used to compute the probabilities to scatter particles off one another, the resulting answers were infinite. In itself, this was not new: the same feature is encountered throughout the Standard Model. However, no one regarded the Standard Model as fundamental. In the

Standard Model there is, as described in chapter 3, a conceptual understanding for why these infinities occur. The Standard Model is not the correct theory of nature on the smallest distance scales, and so it is natural that it produces infinities when the calculations are extended to these scales. In the correct theory of nature at the smallest scales, these infinities should go away – but they were not going away in supergravity.

In addressing any major open problem, calculation is normally not enough. Hard calculations are never enough: it is too easy to go wrong and create the illusion of a result. What convinces is calculation combined with physical reasoning. What caused excitement was that in string theory the same calculations gave finite answers – *and there was a good physics argument why they should give finite answers.* This physical argument came from the fact that strings are strings: they are not points. The finite size of the string smears out all interactions, and the infinities which are caused by point-like constituents go away when the constituents are string-like. This explained *why* the string calculations were getting finite answers – and it also gave a motivational reason for why the calculations should always give finite answers.

Of course, this is not a logical proof. The Nobel Laureate Steven Weinberg expressed this in 1988 as follows:

> There are hand-waving arguments that are not at all rigorous, perhaps not entirely convincing, that the theory *ought* to be finite to all orders. Then when one works out how it really looks in the lowest order of perturbation theory, one finds that those hand-waving arguments really work ... With superstring theory I think finiteness is a reasonable guess. I'd be more surprised than not if it weren't finite.

Despite this, it was not these finiteness results that would trigger the rush towards string theory.

The second problem with supergravity was the problem of consistency. A subtle aspect of quantum mechanics is the existence of anomalies. Anomalies are effects that can render a theory entirely inconsistent as a quantum mechanical theory – even though the same theory makes perfect sense in classical physics. While anomalies are subtle, their effects are drastic: a theory that fails the anomaly check is like the applicant for intelligence work who fails the security check. It is the end: no other extraordinary features or properties can ever be enough to compensate for this failure, and such a theory is inconsistent and irredeemable.[7]

A key feature of anomalies is also that they are a low-energy effect rather than a high-energy effect. To determine whether a theory is anomalous, it is sufficient to know only the spectrum of light particles, with no need to know

---

[7]In case anyone is wondering, the Standard Model is free of anomalies. This fact essentially guaranteed the discovery of the top quark, which after many years of searching was found in 1995 at Fermilab outside Chicago. Without this particle, the Standard Model would have been inconsistent – so it was not a shock when it was found.

how the theory behaves at the highest energies. A thorough study of the effects of anomalies, with respect to both gravitational and other forces, was performed in 1983 by Luis Alvarez-Gaumé and Edward Witten. They found that the most interesting supergravity theories, the ones with the potential to incorporate both gravity and forces resembling those of the Standard Model, were anomalous – and therefore incompatible with quantum mechanics.

Influential physicists were then politely sceptical that string theory could improve on supergravity. String theory appeared to modify supergravity theories, but only at high energies. It was the low-energy properties of supergravity theories, apparently shared by string theory, that seemed to render them inconsistent.

It would turn out to be the success of string theory in addressing this problem that caused the change in attitude. The reason for this impact was not purely the result by itself – it was also that the result showed a significant incompleteness in a quantitative calculation by some of the most respected theoretical physicists in the world. It was the computation of anomaly cancellation in superstring theory by Michael Green and John Schwarz in the autumn of 1984 that propelled string theory into the active brain of theoretical physics – a position that it has never left. The calculation they performed did not find any error in the calculations of Alvarez-Gaumé and Witten. What they did find was that in string theory this calculation always had an extra term – a term that had the precise effect of cancelling the supergravity anomalies that had been found by Alvarez-Gaumé and Witten. The supergravity that was the limit of string theory was not *quite* the same. String-flavoured supergravity had one crucial extra term in it – and this precisely cancelled the anomalies. The same extra term then also served to remove the fatal anomalies in all other calculations as well.

It could in principle have been possible to add this term into supergravity in an *ad hoc* way, although this possibility had not been anticipated in advance. However, it was the way that this term just dropped out of string theory that gave the structure both plausibility and consistency. At the point where string theory had to fail and be marked as inconsistent, it succeeded – and in a way that had not been foreseen in advance as possible.

At the time there was another aspect to their result that appeared striking. Previous calculations had shown that superstring theory worked in ten, and precisely ten, spacetime dimensions. The anomaly cancellation equations of Green and Schwarz also revealed something similar. The cancellations that they found worked only for a single choice of the forces present in the ten-dimensional string theory.[8] It seemed that the more that was learned about string theory, the more unique its structure appeared.

While thunderstorms raged amidst the Colorado mountains outside, Green and Schwarz developed their result at the Aspen Center for Physics, announcing it in the autumn of 1984. This result rapidly reached Edward

---

[8]More precisely, the cancellation occurred only for an $SO(32)$ gauge group.

Witten in Princeton, then in the early stages of his period as the single person of maximal influence in theoretical particle physics, a period that would endure for the next twenty years or so. He embraced string theory. Everyone else followed, and the backwater became a torrent.

In 1984 Ronald Reagan swept to a landslide victory against Walter Mondale, famine raged in Ethiopia while in Britain the miners' strike began its long, slow, tragic collapse. What was string theory? It had just gone from a failed theory of the strong interactions to the number one hottest topic in theoretical particle physics.

5.5   THE LIMELIGHT

The five years from 1984 onwards would see a huge increase in the number of physicists working on string theory. However this is not a history chapter for the sake of history. It is rather to address the question of what 'string theory' *is*, and part of the answer involves knowing how the understanding of the subject has evolved through time – and what string theory was in this period was very definitely still the theory of quantum mechanical relativistic strings. String theory was a theory of strings, with aspirations towards being the correct quantum mechanical account of gravity and the other forces. This wish had been vocalised in 1974; while the rate of work increased, research in the subject in the mid-to-late 1980s was still guided by this earlier goal.

Much effort went into developing the most efficient and economical tools to analyse strings, as well as learning the mathematics that went along with them. The powerful physical and mathematical technology that was thrown at the subject in this period was however intellectually the continuation of work originating in the early 1970s. The number of people working on the subject increased dramatically: the general theme of research remained the same.

What was the big intellectual picture of the subject in this period? There were five consistent string theories known. There were the two type II theories – type IIA and type IIB – that had been formulated by Michael Green and John Schwarz in 1981. There was also the type I theory: this was the one in which Green and Schwarz had performed their by now famous anomaly cancellation calculations. In this paper however, Green and Schwarz had also observed that there was in principle a second solution to the anomaly cancellation equations. One solution had corresponded to the existing type I string theory. What did the other correspond to? Green and Schwarz suspected that it represented another, as yet unknown, string theory, and they speculated as much in their paper. They were right. This was the heterotic string, found in January 1985 by the so-called Princeton string quartet of David Gross, Jeffrey Harvey, Emil Martinec and Ryan Rohm; the name arose as it was a (consistent) hybrid of a bosonic string and a superstring.

The heterotic string theory was also found to come in two variants. One variant was similar to the existing type I theory – in particular, it contained the same kinds of forces. The other represented the 'other' solution to the anomaly cancellation equations: the $E_8 \times E_8$ heterotic string.

Five string theories, each apparently consistent, were known. No fundamental principle was known that could select between the five. What to do next? There was at least a practical way of choosing. Some of the theories – the two type II theories – appeared to exist as purely gravitational theories. They were theories of gravity, and that was all. There did not seem any prospect of connecting these type II theories into observed particle physics. The other theories – the type I and heterotic theories – contained a particle content that, viewed from a distance in the fog and without glasses, resembled that of the Standard Model. One theory in particular, the so-called heterotic $E_8 \times E_8$ theory, seemed particularly promising in this respect. The forces it contained were more elaborate versions of those present in the Grand Unified Theories that were thought to lie behind the forces of the Standard Model.

The heterotic $E_8 \times E_8$ theory was still a theory defined in ten dimensions. To make a model of four-dimensional physics, it was necessary to 'compactify' the theory. This involved curling up six of the extra dimensions to be small and unobservable, so that their existence would have escaped our notice until now. Once this is done, this produces a theory that looks like a four-dimensional one – with particles, forces and interactions that depend on exactly how the curling up is done. It was then realised that if the extra dimensions had the geometry of what is called a Calabi-Yau space, the four-dimensional theory and its particle content would make a passable imitation of the supersymmetric grand unified theories that were then in vogue.

Having taken a decade's leave of absence from particle physics, this suddenly offered a re-entry route for string theory back into the traditional enterprise of physics: understanding, explaining and predicting experimental data.

The structure of ten-dimensional string theories had been uniquely constrained through a variety of consistency conditions. The number of dimensions had been fixed by consistency, and the allowed types of forces present had also been fixed by consistency.

It was hoped that similar consistency conditions would apply on the reduction from ten to four dimensions. Amidst a wave of exuberance, it was thought that string theory may be on the verge of explaining the forces, masses and particle content of the Standard Model. There was no actual calculation that suggested this, but with a rush of new results coming in, it seemed that it might be only a matter of months before another new result would make it possible to compute the mass of the electron from first principles. This sense that a major understanding of the Standard Model might be just around the next corner can be felt in the abstract of one of the first papers on superstring models of particle physics from the spring of 1985, entitled *Vacuum Configurations of Superstrings*, and written by Philip Candelas, Gary Horowitz, Andrew Strominger and Edward Witten.

> We study candidate vacuum configurations in . . . $E_8 \times E_8$ supergravity and superstring theory that have unbroken $N = 1$ super-

symmetry in four dimensions. This condition permits only a few possibilities, all of which have vanishing cosmological constant.

Today there are at least 473 800 776 such candidate possibilities known.

'String theory' was suddenly no longer just a remote theory of quantum gravity; it now appeared that it might also explain the structure of particle physics. It was a heady period when the frontiers of the subject were rapidly advancing, and no one knew where this advance would stop.

Throughout this euphoric period, there were some important known knowns – or so it was thought. Two of these knowns were that the correct number of dimensions was ten and that the correct objects to think about were strings. In particular, the supergravity theories that had been studied before were merely classical limits of the full string theory. It was felt that real physicists should study the full theory and not the old approximations. Older theories such as eleven-dimensional supergravity, together with their membranes with two extended spatial dimensions, were regarded as accidental and erroneous artefacts which could now be discarded.

There was one large, clear, known unknown in this period. We have already seen in chapter 3 that quantum field theory is easiest in a certain limit: the weak coupling limit. This is the limit in which particles interact only very weakly with each other. Answers are obtained by computing successive terms in a series and then summing these terms. In this limit, the later terms are merely small corrections to the first terms. While useful for precision work, they are not needed to obtain a qualitatively correct answer. This perturbative limit is the easiest limit for calculational purposes, but is not the full story.

String theory throughout the 1980s was a subject also defined in this perturbative limit. There was a single coupling, the string coupling, which was the precise analogue of the couplings of quantum field theory. The only known way to calculate was through an approximation series in this coupling, valid when the coupling was small. It was known how to calculate the first terms in the perturbative expansion, but even there the calculational difficulties rapidly grew insuperable. It was entirely unclear what lay beyond the perturbative limit, or how any kind of reliable calculational approach could be found in this regime. It was clearly desirable to move beyond perturbation theory, in the same way it was clearly desirable to end poverty and bring about world peace. The only minor issue was how. The strength of physics is in calculation, and calculations are restricted to what you can actually calculate.

Finally, there were the unknown unknowns. The radical changes in perspective experienced between 1968 and 1985 made it seem unlikely that all good ideas had been found. Edward Witten spoke about this in 1986:

> One has to remember that string theory ... is already eighteen years old, and looking back into the past we can see that ten or fifteen years ago there was a long road ahead, a lot of things that weren't known that had to be known, and it's probably still true today.

The words would be prescient – and it would turn out that it was the deprecated and discarded branes of supergravity that would be key to moving beyond the perturbative limit.

5.6 THE THEORY FORMERLY KNOWN AS STRINGS

With this attitude, string theory moved into the 1990s. After the great rush of work in the five years following 1984, the level of activity was reducing. It was a time for consolidation. It was a time for the exploitation of previous results rather than the development of new ones.

It was also a time when excitement was building elsewhere. Stung by the discovery of the W^+, W^- and Z^0 bosons at CERN in 1983 – as the *New York Times* editorial put it, 'Europe 3 United States – not even Z-zero' – the Land of the Free and the Home of the Brave had embarked on an ambitious undertaking to take back, for good and for ever, leadership of experimental particle physics. A site in Texas had been selected for the construction of the Superconducting SuperCollider. The plan was to construct an 87 kilometre tunnel and fill it with superconducting magnets, generating at Waxahachie National Laboratory the largest and most powerful accelerator complex in the world. The project was American. The scale was Texan. It was deliberately on a size that CERN, whose own plans for the Large Hadron Collider were based on reusing an existing tunnel, could not compete with. The project was approved and construction started. Gigantic boring machines excavated kilometre after kilometre of tunnel deep underground the Texan soil. Had it been built, the SuperCollider would have been twenty times more powerful than any collider then existing, and even three times more powerful than the Large Hadron Collider following its 2015 energy upgrade. The vista of experimental particle physics would have been thrown out to the horizon, and then again.

It was not built. On November 3rd, 1992, William Jefferson Clinton was elected the forty-second president of the United States, replacing George H. W. Bush. The Superconducting SuperCollider had been conceived in the Reagan years and had also enjoyed the strong support of the one-time Texas congressman Bush. It was a project associated with the previous Republican administration and, by 1993, a project facing questions over management quality and budget escalation. With the United States in recession, and Democratic control of the Presidency, the Senate and the House of Representatives, the Superconducting Super Collider was cancelled on October 21st, 1993. This cancellation represented a blow that experimental particle physics in the United States has never recovered from. Current leadership in the subject sits, without question, at the outskirts of Geneva. In the long-term planning for the subject, possible locations for future colliders include CERN, China and Japan – but not the United States.

The reason for this mild digression is to explain some of the background reasons that caused so many young theorists to work on string theory in the period I am about to describe. In part, this was driven by a set of important breakthroughs that occurred around 1995. However it would be wrong to ignore the other external circumstances. The short-to-medium term future for experimental high-energy physics had just turned far bleaker, and it was natural that work tending towards predicting and understanding data became less attractive to any young would-be hotshot.[9]

The prevailing view of string theory changed in 1995 as problems that had appeared far-off and intractable were suddenly revealed to be nearby and soluble. This occurred not through any one single discovery, but through a new perspective on a variety of existing known results. The traditional dating of this change is to a talk by Edward Witten at the Strings '95 conference at the University of Southern California on March 14th, 1995: both the talk and Witten's subsequent paper summarising it brought a complete change of outlook. However, it was not that everything Witten said was new; and it was not that everything Witten talked about was specifically his own work. The rapid and dramatic change in outlook was due both to Witten's synthesis of disparate results into a clear single picture and to the commanding personal influence that Witten had in the field.

Let me try to summarise briefly the changes of 1995. First, it was realised that the five different string theories then known were all related. They were indeed more than related – they were connected. It was possible to interpolate continuously from one theory to another. The five theories were merely individual parts on a continuously connected web in theory space. As mentioned earlier, in each theory the strength of the interactions is parametrised by something called the *string coupling*. The string coupling is the analogue in string theory of the coupling constants in quantum field theory: it determines how likely it is that two strings will interact, and how likely it is that a single string can split into two distinct strings. The calculations of the 1980s were performed at weak coupling, a limit where these interactions occur rarely. The question was: what happened at strong coupling, where the techniques of the 1980s all broke down?

The surprising answer of 1995 was that the theory at strong coupling simply turned into either the same theory at weak coupling, viewed from a different light, or into a different theory – but also at weak coupling and in the regime where it was easy to calculate. These are examples of the dualities described in chapter 3.

What is meant by the same theory at weak coupling? It means exactly the same theory, with exactly the same equations – but with some of the labels changed. For example, suppose the original theory, defined at weak coupling, had equations involving two terms 'B_2' and 'C_2'. The equations for this theory at strong coupling are exactly the same, but wherever we wrote

---

[9] In the years leading up to the turn-on of the Large Hadron Collider, exactly the opposite trend was seen. String theory and other more formal topics became unpopular, and the physics of colliders was the hot topic.

'B_2', we now write 'C_2'. If we relabel 'B' as 'C' and 'C' as 'B', we are back where we started.

We can imagine moving along a line starting from weak string coupling (infinitesimally close to zero) all the way to strong string coupling (where the coupling becomes formally infinite). The point we end at is in no way exotic. In some cases, the end of the line is the same as its start: it resembles the Circle Line rather than the District Line. This is indeed actually the case for what is called type IIB string theory – the limit of this theory at strong string coupling is exactly the same as the theory at weak coupling, except with some labels interchanged.

In other cases, the strong coupling limit of one theory turns out to be a different theory at weak coupling. This applied for the type I SO(32) theory and the heterotic SO(32) theory: one theory taken to strong coupling turned out to be the other theory at weak coupling. Although these two theories had appeared quite distinct – involving very different particles and interactions – they emerged as the same theory, simply in different regimes. The part of each theory that appeared impenetrable was simply the other theory at its most accessible point.

These results did not just materialise out of the mist. These dualities extended previous examples of the phenomenon. A related question was the behaviour of theories at small and large values of the radii of the extra dimensions. The physical conditions here seem very different: a cylinder may remain a cylinder irrespective of its size, but a large cylinder is surely a different cylinder than a small one. However, there are again dualities that relate small and large radius: for example, the type IIA theory on a cylinder R times larger than the length of a string is exactly the same theory as the type IIB theory on a cylinder R times smaller than the length of a string. This duality is called T-duality, and it was already a well-known result that had been discovered back in the 1980s.

The idea of *S-duality* – the statement that a theory at strong coupling could be identical to a theory at weak coupling – went back to 1977 work on quantum field theory by Claus Montonen and David Olive. Olive had been one of the pioneers of string theory, but like many others had moved on to other topics during the late 1970s. In 1990 the idea that superstrings exhibited strong-weak coupling duality had been put forward by Anamaria Font, Luis Ibáñez, Dieter Lüst and Fernando Quevedo – although within the ultimately incorrect context of the heterotic string.

However, how could these ideas ever be checked, given the difficulties of computing at strong coupling? A key step was to find properties of the theory that were, in a technical sense, 'protected': they could be calculated at weak coupling and the result extrapolated to strong coupling. In 1994 Ashoke Sen, an Indian physicist as brilliant as he is modest, found an example of such an object. S-duality required this object, which had both electric and magnetic charge, to exist – and Sen was able to show it did exist, as required by duality. The claim of an existence of a larger duality called U-duality, unifying previous examples, had also been put forward six months before Witten's talk by Chris Hull of Queen Mary and Westfield College in London and Paul Townsend of Cambridge.

As in 1995 Witten brought both his own insights and all these other results together, the picture he outlined included a further shock to the system, in that the answer to 'What is string theory?' was suddenly no longer 'First, a theory of strings.' The answer was now 'a limit of M-theory'. What was M-theory? It was the unifying theory of which all string theories were simply small parts. The different string theories became seen as different limits of one object. That one object was M-theory – and whatever M-theory was, the understood string theories were just one limit of it.[10]

One of the most surprising aspects of this was that there was another limit of M-theory involving a classical gravity theory in eleven – not ten – spacetime dimensions. This was the eleven-dimensional supergravity theory that had first been constructed in 1978. This limit contained no strings. Instead, the basic extended excitations were not strings, but rather objects with two spatial dimensions – membranes. The theory of these membranes had been worked out by Eric Bergshoeff, Ergin Sezgin and Paul Townsend in 1987. The eleven-dimensional limit was accessed by starting with the ten-dimensional type IIA theory and making its coupling strong. In this limit, the coupling itself morphs into an extra spatial dimension – another realisation of the dynamical and mutable nature of stringy geometry.[11]

From this new perspective, string theory became at once a smaller and a larger subject. It was smaller because the previously distinct string theories were now part of a single whole: there was only one 'thing' to be understood, and everything else was just a different limit of this single thing. It was a larger subject because the underlying equations of M-theory were, and are, unknown. M-theory is known from its boundaries. The Picts and the Numidians could infer the existence of Rome from their battles with its legions, but could never know that it has seven hills. M-theory is in essence defined as the entity whose different limits give either the various known string theories or eleven-dimensional supergravity: but its underlying equations are not known.

The connection of string theory to eleven-dimensional supergravity was not unprecedented either. Indeed, it was as early as 1987 that Michael Duff, Paul Howe, Takeo Inami and Kelly Stelle, all then at CERN, had pointed out that you could make a ten-dimensional type IIA superstring by taking a membrane in eleven dimensions and viewing it in the limit where the eleventh dimension was very small.[12] For those who had continued working on

---

[10]What does the 'M' in M-theory stand for? It is a bit like the 'S' in Harry S. Truman – it is polymorphous and can stand for more than one word, for example 'magic', 'mother', 'mystery' or 'membrane' according to taste.

[11]At first sight this may seem hard to get one's head around; it remains so at twenty-first sight.

[12]When I first read this paper I was quite shocked by its existence; according to the supposed history of string theory that I had 'learned', such a paper could not have been written for almost another decade.

supergravity during the 1980s, and even more so for those who worked on eleven-dimensional supergravity in this period, this was a time of intellectual vindication. For a decade they had had to face a certain amount of condescension for not making the shift to strings – Michael Duff reports having to deal with comments such as:

I want to cover up my ears every time I hear the word 'membrane'.

In 1995 then, ideas that had been mostly ignored suddenly came to prominence: these different pre-existing elements were drawn together by Witten into a new picture of string theory and how the subject worked.

The membranes were also not confined to supergravity and were not confined to eleven dimensions. Membrane solutions had also existed in the ten-dimensional supergravity theories that were the classical limits of string theory. However, whether these solutions meant anything in string theory was highly unclear. Soon after Witten's talk, Joe Polchinski of the University of Texas showed that membranes also played a crucial role in string theory (under the name of D-branes). Brane solutions had already been known within the supergravity theories that existed as classical limits of string theory. Polchinski and others had also studied the role of branes within string theory, where they appeared to be exotic objects with no simple classical description. Polchinski was able to join these descriptions up. He showed that the supergravity branes admitted an extremely simple description in string theory – and this description also allowed plentiful calculations to be done.

It was soon realised that as the coupling strength was dialled from weak to strong, the string states present at weak coupling turned into the brane states of strong coupling – and vice-versa. This implied a conceptual rethink, as 'fundamental' string states morphed smoothly into brane states. What this showed was that, even in string theory, strings were ultimately no more fundamental than branes. The foundational nature of one-dimensional extended objects – strings – was only an illusion coming from the most calculationally accessible regions.

Let us then return to the question of 'What is string theory?' and ask how this would be viewed in 1996. The answer would have been that the five string theories that had been identified in the 1980s were all really different manifestations of a single underlying object. The underlying theory was called M-theory, but like a great desert it was known only from its boundaries. There were no known fundamental defining equations of M-theory, and strings entered M-theory only in certain limits. The differences between the string theories of the 1980s – the type I and type II strings, and the heterotic string – were like the differences between steam, water, ice and snow. While very different, they all arise from molecules of water. The different string theories were not fundamentally different *theories* – they were instead different limits of the same underlying object. Although converting an iceberg to superheated steam would require an enormous input of energy, it is in principle possible in a way that converting the iceberg to laughing gas is not. Likewise, although the energies required to move the universe from one 'phase' of M-theory to another phase would be unimaginably large, it could in principle be done. Such a change would represent a dramatic alteration in the affairs of the universe, but it would not modify the fundamental laws under which it operated.

String theory in 1996 was then no longer a theory of strings and no longer a theory in ten dimensions. What had not changed, however, was the view as to what fundamental problem 'string theory' was trying to address. String theory might now be called M-theory, and might now appear as a far richer theoretical structure than had previously been dreamed of – but its *raison d'être* was still as a possible theory of nature on the smallest possible distance scales. The form of the answer had changed: the question it was answering had not. Despite all that had been found, the style of work in this period was fundamentally similar in spirit to that of the late 1970s and early 1980s. It was motivated by and addressed the questions 'What is this theory and what is its internal structure?' and not 'What data does it explain and what does it tell us about experiment?'

String theory had started in 1968 as a theory of strong interactions and ended up in 1996 as M-theory. If these twenty-eight years had been a journey of abstraction from explaining data concerning the strong force to formal properties of the one true theory of quantum gravity, the period since then has involved travel in the opposite direction. Ideas and techniques that were originally spawned as offshoots are now used in their own right to attack many different problems. Far more people now work on string theory either as a framework for ideas or as a source of calculational techniques than on the apparently canonical question of quantum gravity. Indeed, in an amusing turn of the wheel one application of 'string theory' is now, once again, to compute amplitudes for scattering via the strong force.

We now turn to these more modern attitudes to string theory.

CHAPTER 6

What Is String Theory?

Although for many years the story of string theory had been viewed as an attempt to construct a fundamental theory of nature, few now see it solely in this light. The year on which this story pivots is 1997. Up until then, string theory was broadly felt to belong on the top floor of an intellectual tower. Its practitioners regarded it as inhabiting the penthouse suite of ideas: above quantum mechanics, above general relativity and above quantum field theory. Quantum field theory was regarded as something that arose in certain limits of string theory when stringy effects were removed. The converse was not true – strings, and quantum gravity, did not arise from limits of quantum field theory. Indeed, even classical gravity was absent from quantum field theory. The relationship of string theory to quantum field theory was like quantum mechanics to classical mechanics – an upgrade that, while backwards compatible, also added many new features.

While the change was gradual, a principal source of the transformation in attitude was the discovery in 1997 by Juan Maldacena of what is called the AdS/CFT correspondence.

6.1   QUANTUM FIELD THEORY BY ANY OTHER NAME

The AdS/CFT correspondence will be the principal subject of chapter 8, but I shall give here a brief summary. The correspondence is a duality between a gravitational theory and a gauge theory, and it was the last really major result coming from the period around 1995.

In particular, Maldacena claimed an absolute equivalence between particular quantum field theories and particular solutions of string theory. As a duality, this was in one sense similar to the earlier discoveries of the 1990s. It was a statement that two objects were identical, except expressed in a very different language. However, it was also a very different form of duality.

In the previous examples, the theories related were theories at the same level. These dualities related quantum field theories to quantum field theories, or string theories to string theories. However different the details appeared, they involved objects on the same footing. Maldacena's claim involved a string theory on one side and a quantum field theory on the other. The former was a theory of gravity, formulated in terms of extended objects. The latter was a theory of point-like particles. This time, the two sides of the duality involved different types of theory.

This result certainly offers deep insights into quantum gravity, and so could be seen as part of the ongoing string theory research program at that time. It demonstrates the holographic nature of gravity – the statement that a gravitational theory in D dimensions can be equivalent to a non-gravitational theory in $(D-1)$ dimensions. While it is true that Maldacena's correspondence is not universally applicable, and in particular only applies for cases when the energy associated to the vacuum of space is negative, it is general enough to provide powerful conceptual and calculational insights into quantum gravity.

However, this same correspondence spectacularly muddles up the issues of what is fundamental and what is emergent. While quantum field theory is present within string theory, it turns out that for many interesting cases string theory is also present within quantum field theory. In some cases, string theory just is quantum field theory. Physics that had been thought to be the exclusive preserve of gravity was now arising in a subject that was not meant to have any knowledge of gravity.

In the winter of 1997, Britain's popular new Labour prime minister, Tony Blair, had been in power for six months with a landslide majority, while the territories of Hong Kong had just returned to the governance of the People's Republic of China. In Moscow, one Vladimir Putin had recently joined the junior staff of President Boris Yeltsin, while in Connecticut the hedge fund Long Term Capital Management was on its way to total and spectacular collapse. In the town of Palo Alto in California, Larry Page and Sergey Brin had just registered the domain name Google.com.

At this time another answer to the question 'What is string theory?' had just emerged, and that answer was 'quantum field theory'.

When Maldacena's original paper appeared on the arXiv, it was unclear how things would turn out. It came at a time when many exciting results had just been published, and there were many promising directions. The paper was rapidly recognised as important though, and large numbers of those who were interested in either string theory or quantum field theory jumped on it. Indeed, it took less than two years from the original paper for the first major review article to appear, an article which itself went on to become a standard reference from which many students would end up learning the subject.

The early works focused on elucidation, clarification and evaluation of the correspondence. How did the correspondence work in practice? Which calculation in string theory corresponded to which calculation in field theory? Did they indeed match? The performance of such computations is the regular work of theoretical physics, and these checks have been used to provide greater or lesser levels of employment for almost twenty years now. As will be described in chapter 8, these checks now involve exceedingly intricate results, and there can be little serious doubt that the correspondence is correct.

A large practical change triggered by Maldacena's discovery arose from the fact that it also opened the door to a more pragmatic role for string theory as a calculational tool. If in certain cases string theory is equivalent to quantum field theory, then for these cases calculations in string theory are also calculations in quantum field theory. On consulting the dictionary provided by the correspondence, any calculation performed in string theory also has a meaning within quantum field theory. If you calculate in one, you calculate in the other. The value of this comes from the fact that under a duality, easy calculations in one theory are related to hard calculations in the other theory. This allowed string theory to be turned into a tool for doing hard computations in quantum field theory.

During the first decade of the new millennium, this activity was probably larger than any other occurring under the name of string theory. It came in several forms. One part was devoted to understanding model examples of quantum field theories: field theories that did not and could not describe nature, but which served as tractable examples where these techniques could be deployed in all their power. These are typically theories with supersymmetry. For this purpose, some supersymmetry was good; more was better. The more supersymmetry, the more control and the more ability to deploy the correspondence with rapier-like precision.

Another part was devoted to making AdS/CFT useful for field theories with some applicability to the real world. The best way to define 'applicable to the real world' was certainly not clear at first. The most obvious target was the theory of the strong force, quantum chromodynamics. The canonical examples of the AdS/CFT correspondence applied to theories which are cousins of the actual strong force, except with far more supersymmetry. If it were possible to lose the supersymmetry while retaining the correspondence, one might obtain interesting results for the actual strong force. This research, aiming to reproduce the properties of the actual strong force, has now been carried on for over a decade, with more or less degree of rigour and less or more degree of success.

In the large review of 1999, reproducing the actual strong force was the only 'real world' application mentioned. Since then, as we shall see in chapter 8, the number of applications of the AdS/CFT correspondence has proliferated. We briefly mention them here: one application is to understand what happens when heavy nuclei such as gold or lead are collided with one another, while another is to provide a new perspective on systems with both strong interactions and very many atoms. While both applications draw on string theory and explicitly use the AdS/CFT correspondence, it is also clear that they do not rely on quantum gravity. In these applications, string theory may be a tool and it may be a framework – but it is not serving as a fundamental theory of nature.

One effect of Maldacena's result was then to make string theory less special. It diluted the attitude that string theory was in some sense 'better' than quantum field theory. If quantum field theory could be string theory, and string theory could be quantum field theory, then it was harder to argue with a straight face that string theory was more fundamental. While string theory was still a broader framework than quantum field theory, the sense of difference was reduced. It also greatly increased the number of ways in which string theory could be treated as a tool, compatible with an agnostic attitude to its status as the fundamental theory of this world.

6.2   THE MISANTHROPIC LANDSCAPE

In parallel with this, another topic which has seen a significant change over this period is the attitude towards string theory as a theory of the world on the smallest scales. String theory is naturally defined in ten, or possibly eleven, dimensions, which is in both cases a number much larger than four. The only way to obtain a world that looks four-dimensional is by compactification – curling up the extra, unwanted dimensions. The four-dimensional *mappa mundi* is then determined by the shape and geometry of the extra dimensions. Predictive statements about our world require knowledge of the extra-dimensional geometry.

What determines this geometry? Broadly, there are two options. Either the dynamics of the theory determine the geometry uniquely, in which case it can take only one form, or they do not, in which case there are many options. If the former case were to hold, only one consistent choice of geometry would exist. This would lead to a uniquely predictive framework for four-dimensional physics. Just as quantum field theory predicts the mass of the positron to be identical to that of the electron, string theory would lead to a unique prediction for not only the electron mass, but all aspects of the Standard Model.

On the other hand, if the latter case holds, string theory would have no more ability to predict the exact extra-dimensional geometry than Newton's laws have to predict the exact number of planets in the solar system.

The general attitude during the 1980s had been that it was the first case that would ultimately hold. String theory was not understood in many ways. We saw in the previous chapter some of the many unknowns then present. At the lowest levels of approximation, there did seem to be many consistent ways of curling up the extra dimensions. However, it was thought that these would melt away once the theory was better understood, leaving only a single allowed option.

There was certainly no proof that this would happen. However, this attitude would have been moulded by the previous history of string theory, where a plenitude of choices in both spacetime dimensionality and the allowed forces had been shrunk to a single option by subtle and intricate consistency conditions. It was reasonable to believe that as understanding developed further, something similar would happen for compactifications. If it did, then string theory would lead to only one set of forces, one set of particles and one set of interactions – and possibly, an argument for why the Standard Model had to be the way it was.

Over the last fifteen years this attitude has essentially disappeared. There is now a widespread agreement that string theory provides no unique route from ten to four dimensions. Several factors have influenced this change in sentiment.

While during the 1980s, there was great uncertainty as to what would happen at strong coupling, now there is a far greater understanding, in particular via branes, of effects in string theory that go beyond weak coupling. For certain special systems, exact solutions have been found for both quantum field theory and string theory. With the domestication of the dragons and sea monsters, the idea that consistency conditions will select a single solution now seems less plausible. For cases with large amounts of supersymmetry, fully exact solutions appear to exist, with nothing wrong with them. Even for more 'realistic' solutions with small or zero amounts of supersymmetry, no gaping holes have appeared in the approximate arguments that suggest the existence of many consistent ways of going from ten to four dimensions.

As with any broad change in attitude, we should be careful not to attribute it solely to a single result or to the effect of one prominent person. From the very first days of compactified string theory, there have been those who expected string theory to produce a large number of solutions. Today, there remain some who hope that there still exists a unique, so far unidentified, selection principle that will select a single way of going from ten to four dimensions.

The fact that the fundamental equations of a theory do not have a unique solution is not confined to string theory. Maxwell's equations of electromagnetism allow for both radio waves and visible light. These equations constrain the allowed form of electric and magnetic fields in a volume of space, but they will not tell you what structure you actually find. In a similar way, Newton's laws of gravity do not allow us to deduce that the earth has one moon but Jupiter over sixty.

So, how many correct ways are there to go from ten to four dimensions in string theory? The apparent answer is infinity. There are a large number of exact supersymmetric solutions which have continuous parameters – in particular, type II strings on Calabi-Yau geometries. As these parameters are continuous, they can take an infinite set of values. With an infinite set of choices, there are an infinite number of solutions.[1]

---

[1]From a more technical point of view, the reason these solutions are exact is because they preserve $N = 2$ supersymmetry. This additional supersymmetry controls corrections to such an extent that the results are guaranteed to be exact. The heterotic string on a Calabi-Yau preserves $N = 1$ supersymmetry – which, while useful, is not enough to guarantee an exact solution.

For all these riches, not a single one of this infinity of consistent compactifications can ever describe our world. This is because every last one of them preserves exact unbroken supersymmetry, and we know that supersymmetry is not an exact symmetry of our world. Exact supersymmetry would imply that the electron had a bosonic partner with the same mass and the same charge – but no such particle exists.

To describe our world, supersymmetry must be a broken symmetry, and the compactification must fail to preserve supersymmetry. How many solutions then exist? The lack of exact supersymmetry makes the calculations harder and more vulnerable to error, but as the understanding of string theory grew, so did confidence in these calculations.

It should be emphasised now that there remains a minority who think that these calculations miss something deep and are not reliable. However, if the calculations are trustworthy, one can ask how many ways there are of going from ten dimensions to four dimensions. A few years after the millennium, a crude estimate was made for this, resulting in a number of 10^{500}. This number stuck, and went viral. This large number of apparent solutions of string theory is now often called the string theory landscape.

Whatever its merits, let us stick with this number. This number would say that, whatever the status of string theory in ten dimensions, there are roughly 10^{500} consistent ways of curling up six of the dimensions to turn the theory into a four-dimensional one. This is a vast – almost inconceivably vast – number. Does this mean that string theory is dead, or at least useless, as a theory of nature? If there are 10^{500} ways of going from ten to four dimensions, does this render it useless as a part of science?

While the number 10^{500} sounds gargantuan, reflection shows that this argument is a little silly. Its silliness may be seen through the more familiar issue of genetics and human reproduction. Every one of us has a genetic code inherited, neglecting occasional mutations, in part from our mother and in part from our father. The number of combinations that could have made up our DNA is vast. Personally, I have two sons, and at the time of writing the younger is two and the older is four. There were far more than 10^{500} possibilities for how our younger son could have turned out. With such a landscape of options, who could ever possibly have predicted the true fact that he looks similar but not identical to his big brother at the same age?

Large numbers may intimidate, but they should not scare. What they should provoke is not flight, but instead reflection on what is the right question to ask. As we have seen, the possible genetic variation in humans is vast beyond count. It is impossible to guess someone's genetic code, either before or after meeting them in person. Nonetheless, despite the enormous number of possible genetic combinations, you are likely to do quite well by predicting that they have ten fingers, do not have blue skin, and have an adult height somewhere between four and a half and seven feet.

In the context of string theory, what are the right and wrong questions? A good example of a wrong question is to ask what the precise geometry of the extra dimensions is. It is not needed and not relevant. You need not know every mutation on someone's genome to know they have brown eyes. A more productive line of enquiry is to ask about the general features that arise and the observational consequences they lead to.

We will explore this direction at greater length in chapter 10. To give one example here, compactifications almost always involve large numbers – hundreds or even thousands – of scalar particles called moduli, which have gravitational-strength interactions. It is sensible to search for the consequences of having large numbers of such particles. For example, one can ask: would the presence of a thousand moduli, as opposed to five, lead to observational effects in string theory models of particle physics and cosmology?

This is a type of question that is productively asked. In asking such questions, there is no claim to be identifying the one true theory of the world. However, it takes string theory seriously, treats it as a framework and asks what the typical properties are that this framework leads to. These properties may be testable in the conventional way, and the fact that this does not uniquely reveal the extra-dimensional geometry should be no more disappointing than the fact that long black hair does not tell us someone's genome.

This is illustrative of the general trend in applications of string theory to particle physics and cosmology. It is no longer expected that string theory will lead to one unique solution for physics in four dimensions. Rather, string theory is more understood as a framework for constructing models, analysing ideas and testing conjectures. If one wants to ask whether some interesting phenomenon, with clear observational consequences, can be realised within a theory of quantum gravity, string theory can be used as a testing ground. It may provide either a proof-of-principle example confirming this, or alternatively an argument suggesting the phenomenon is impossible. At the time of writing, one such example concerns the possible existence of what are called *tensor modes* within the cosmic microwave background.[2]

String theory appears to have many solutions. This is neither unusual nor unprecedented in science. The existence of many solutions makes it important to think clearly about the questions it is profitable to ask. However, this existence of a landscape of solutions is behind one particular topic where speculation and conjecture have far out-run reasonable argument. This is the topic of the observed vacuum energy of the universe, often called the *cosmological constant*, and the theoretical explanation for this phenomenon.

The most striking experimental discovery over the last twenty years has been the discovery that the universe is accelerating. The universe is not only expanding, but the rate of expansion is increasing. Seventy per cent of the energy of the universe lies in the form of a mysterious 'dark energy' or 'cosmological constant'. This dark energy is associated to space itself – it is present wherever space is, and the amount of it is proportional to the

---

[2] At a technical level, these occur when fields undergo what are called trans-Planckian excursions during an epoch of inflation. It is currently a topic of very active debate whether such excursions are allowed by quantum gravity, or whether a no-go result may exist.

volume of space. What makes this energy so mysterious is not its existence – pretty much any theory of particle physics seems to produce such an energy. The mystery comes from its size. The energy is measured at around sixty orders of magnitude smaller than any reasonable estimate; that is, around 0.000000000000000000000000000000000000000000000000000000000001 of what sensible estimates suggest. While the smallness of the cosmological constant had been known for a long time, prior to this discovery it was hoped that there might exist some unknown deep principle that caused the vacuum energy to vanish. No such principle was known, but it was not difficult to conjecture that one might exist. With this possibility removed, the question has to be faced of why the cosmological constant is both non-zero and also so much smaller than all reasonable estimates.

The disagreement between theory and experiment here is so vast that it is almost an insult to the subject. It is an experimental result that openly mocks any claims of theorists to understand the world. Manifestly clear and important, the problem represents a plum target for ambitious theorists who want to prove themselves. It has attracted many, but no solutions have been proposed that have been found satisfactory or generally acceptable.

It is here, it is claimed, that the presence of many solutions of string theory rides to the rescue. The string theory landscape, it is said, can explain why the cosmological constant is small – and the explanation is the anthropic principle.

The anthropic principle is essentially the statement that the answer to some questions is conditioned by the fact that we are here to ask them. If I ask why the density of atoms around me is so vastly greater than almost anywhere else in the universe, a reasonable response is the fact that life cannot exist in empty space. As humans, we cannot live in a vacuum. We can only ask the question on earth, as that is the only place where we can be alive. Our answer then lies in the simple fact that we are here to ask the question in the first place.

Few would find this response unreasonable. However, while the anthropic principle is not vacuous, it can be seductive. It offers the dangers of the open cookie jar at Fat Camp – the soft route of easy temptation. It also encourages a solipsistic attitude to science. For example, I could ask why the Cuban missile crisis did not end in mutual assured destruction, with a nuclear conflagration that destroyed the world. If I felt sufficiently brazen, I could respond that the answer is the anthropic principle. If nuclear war had occurred in 1962, my parents are unlikely to have met a decade later, and I would never have been born in 1981. However, many would feel that the fact that I, Joseph Conlon, am right now contemplating the marvel of my own existence is not a satisfactory explanation for why Kennedy and Khrushchev managed to avoid taking their respective nations to war.

In the context of the cosmological constant, the structure of the anthropic argument is as follows. String theory has many solutions – perhaps 10^{500} or greater. Each leads to a different value of the cosmological constant. Just as our galaxy is a tiny part of a much greater universe, our universe is also a tiny part of a much greater multiverse, in which all these solutions are actually realised. Across this multiverse of 10^{500}, 10^{1000} or 10^{1500} universes, these different solutions will realise all possible values of the cosmological constant. Somewhere in the multiverse, there is a universe with a cosmological constant as small as in our universe, and with the same laws of physics as in our universe.

What, the argument continues, are the conditions for life to form? Life requires many billions of years of evolution. Billions of years of stable conditions can only arise in a universe that does not expand too rapidly, which requires a very small cosmological constant. Therefore, a precondition for the existence of intelligent beings who ask the question, 'Why is the cosmological constant so small?' is a small cosmological constant. As all possible values of the cosmological constant are sampled in the multiverse, the anthropic principle then – it is claimed – explains why the cosmological constant is so small. The answer is us and our existence. We only exist to ask the question in universes with a small cosmological constant.

What is wrong with this argument? It is the sheer utter extravagance of the speculation, uncoupled from either rigorous calculation or experimental test. The argument requires the physical existence of 10^{500} additional universes, none of which we can probe experimentally. The argument also lacks the redeeming precision of cut-and-dried mathematical argument.

It has all the flaws of accounting for the entirety of human history by my own existence. We are all of us the product of many ancestors. If history had taken a different turn, we as individuals would not exist – but it would be the height of egotistic solipsism to assert that my existence represents a decent answer to the question of why Brutus killed Caesar. If we look at the entire observable universe, the anthropic explanation of the cosmological constant requires that this has been replicated 10^{500} times beyond our view, every time with different laws of physics and histories. None of these other universes are observable – and all to explain one single number whose empirical existence was established barely a decade ago!

The history of science shows that certain problems are often not mature for solution at the time they are posed. Instead, they have to be left alone for decades or even centuries before they can be sensibly tackled. The idea of atoms had to wait over two thousand years to move from speculation to understanding. Chemists understood the existence of different elements in the 18th and 19th centuries, and that they had different properties, but it would require the advent of quantum mechanics in the 1920s to understand where the elements came from. Until new tools had appeared, these problems were inaccessible.

The existence of a cosmological constant became generally accepted around the turn of the millennium. There is no reason that it needs to be understood in the lifetimes of our grandchildren, let alone within a few years. The ideas needed may be centuries away. The most serious problem with the anthropic landscape is that it provides a cheap and lazy explanation that does not come from hard calculation and also has no clear experimental test. It sounds exciting, but does not offer lasting sustenance, and may even act as a deterrent against necessary hard work developing new calculational tools.

Of course, this does not mean that the anthropic approach is necessarily wrong. However, the triumph of science has been not because it contains ideas that are not necessarily wrong, but because it contains ideas that are, in some important sense, known to be true: ideas which have either passed experimental test or are glued together by calculation. The anthropic landscape is neither of these. It represents incontinence of speculation joined to constipation of experiment.

Sometimes, the only thing that can be done with an intractable scientific problem is to wait – not for new universes, but for another, doubtless very different, set of tricks. The theory of the cosmological constant may well be one such case.

6.3 FIFTY YEARS ON

The above topics are characteristic of the post-millennial view of string theory, and similar features are seen in other applications of string theory – for example, techniques from string theory have also recently been used to improve the methods for calculating scattering amplitudes for the strong force. Unlike in its earliest years, here string theory is not being proposed as the fundamental theory of the strong force – but it is being used to develop more efficient calculational methods within quantum chromodynamics.

While the details in each case may be different, what is common is a view of string theory as a framework, an assortment of ideas, or a set of tools and concepts – but not as something providing the one true answer to a single question about nature.

In 2015 then, what is string theory? The shortest definition is the consistent theory of quantum relativistic strings – and everything thereby included. Despite all changes, there have been no logical additions to the original ideas. The many discoveries made over the last thirty years were already there, implicitly, in the first string theories written down in the early 1970s, even though no one suspected it at the time. The connections between different string theories, the duality relations between strong and weak coupling, the existence of a hidden eleventh dimension in the theory, the AdS/CFT correspondence, the idea of M-theory – none of these were supernumerary accessories to the original equations. They have not been added in. They were all already present in the equations of the 1970s, veiled and awaiting discovery.

'String theory' now also includes all the useful mathematics, ideas and applications found along the way. The AdS/CFT duality relating gauge and gravitational theories – that is string theory. The mathematics of Calabi-Yau geometries – that is string theory. Techniques for doing computations in quantum field theory – these are string theory. Collections of models for the early universe – these are string theory. The term has become a loose one, encompassing a range of topics that are only tenuously connected to vibrating one-dimensional objects.

It is not that the old ideas have gone away – it is just that string theory is now both a theory and a set of techniques. String theory has become an umbrella term both for a variety of communities with different motivations, interests and methods, and also for the problems they study. The name nowadays is as much that of a social grouping as a description of what these people actually do.

Considered narrowly, string theory remains a candidate theory of quantum gravity – a claim about the laws of nature at the smallest possible distance scales. Considered broadly, string theory is now also an amorphous blob of results, techniques, outlooks, ideas, methodologies and calculations. While these draw on the vision of string theory as a theory of quantum gravity, they do not rely on it.

At the time of writing, the most recent Strings conference occurred in Princeton in July 2014. As part of that conference, several grandees of the subject gave vision talks on matters past, or passing, or to come. One of these talks was given by the recently retired Cambridge professor Michael Green, who had started work on the subject over forty years ago when it was a still viewed as a candidate theory of the strong interactions. One of Green's slides gave his outlook on string theory:

> As time goes by and String Theory evolves, it is more and more apparent that it is not just a
>
> 'Theory of String-like Elementary Particles',
>
> but it is a
>
> 'Magnificent theoretical framework that interrelates a very wide range of topics in physics and mathematics'.
>
> The unpredictable trajectory of String Theory since its inception is part of what makes our subject so exciting and so challenging.

If in 2015 – as the Islamic State of Iraq and the Levant claims to re-establish the Caliphate, the 200-1 outsider Jeremy Corbyn is elected leader of Britain's Labour Party, England dismiss Australia for 60 runs and regain the Ashes, and the United States Supreme Court announces a constitutional right to same-sex marriage – you wish to be a card-carrying string theorist in good standing, there is no requirement to work on quantum gravity – and not even any requirement to work on strings.

In the third part of the book, we shall now look in more detail at the reasons why people do choose to work on string theory, starting with the direct experimental evidence for it.

III

What For?

Direct Experimental Evidence for String Theory

There is no direct experimental evidence for string theory.

Why Strings? Quantum Field Theory

Let us try again. Why do people work on string theory?

'Quantum field theory' is three small words: but three small words that have spawned a bookcase full of textbooks. We have seen in earlier chapters that the Standard Model of particle physics is an example of a quantum field theory. Quantum field theories are in a sense just – just! – the quantum mechanics of theories such as electromagnetism. That is, they do not exist as something additional to ordinary quantum mechanics, but are rather particular examples of quantum mechanical theories. They are the quantum mechanical theories of fields (for example, the electromagnetic field). They operate under the same rules and obey the same equations as any other quantum mechanical theory. However, to categorise quantum field theories simply as examples of quantum mechanical theories is as accurate as categorising *Hamlet* simply as an example of a book. The statement is as wrong as it is possible to be while still being right. Quantum field theories are both highly important and filled with subtleties. As outlined in the introductory chapters, there are indeed sufficient subtleties that these theories spent their first twenty years being viewed as fundamentally flawed, and their second twenty years being regarded as calculational black magic, before a deeper conceptual understanding finally arose.

The Standard Model represents one particular example of a quantum field theory. Of all possible quantum field theories that one can write down, the Standard Model is the only one nature is known to use as a fundamental theory. However, large numbers of physical systems are, in practice, described to very high accuracy by quantum field theories. The behaviour of electrons in metals is described by quantum field theory. The phenomenon of superconductivity is described by quantum field theory. Many, many complicated systems involving vast numbers of interacting particles are best described using quantum field theories.

The Standard Model, in principle, involves an infinite number of components: one for each point in space. Systems involving atoms may have only a million billion billion such degrees of freedom, one for each particle in the system. To a mathematician, a million billion billion is closer to one than infinity – two finite numbers are like any other when compared to infinity. To a physicist, a million billion billion is at least approximately infinite, and quantum field theory can be, and is, used with great success to describe such systems.

The exploration of quantum field theories in their many guises is therefore an active part of modern theoretical physics. It cuts across subject boundaries and is part of the toolbox of many. While the main rationale for this study is that quantum field theories do appear in physics in so many ways, it would be a mistake to think that all this research is performed with the goal of a rapid comparison of theoretical prediction and observational phenomenon. Much research on quantum field theory is instead carried out simply to advance the better understanding of quantum field theory itself, and this certainly includes the study of quantum field theories that are not relevant to describing nature. This may sound paradoxical – why study theories that are definitely known *not* to describe nature? However, it is often the case that the particular is illuminated by placing it in a broader context. You understand a child better by knowing his or her parents and relatives; you understand humanity better by knowing the evolutionary tree of other closely connected mammals.

Quantum field theories often belong to continuous families, labelled by the values of the parameters that enter them. For example, in the quantum theory of electromagnetism one such continuous parameter is the mass of the electron. If this were different, the quantum mechanics of atoms would also be different – and so also would be the chemistry and biology derived from them. In principle, every possible value for the mass of the electron corresponds to a different quantum field theory. In the real world, only one of these quantum field theories is realised – the electron mass is what it is, and theories with the 'wrong' electron mass are not correct and do not describe nature. However, it is more than useful to regard the 'right' theory as part of a family of different theories with different values for the electron mass.

Understanding quantum field theory as a whole is also important because the individual quantum field theories that are most important – for example, the Standard Model – may be quite complex. It is hard to calculate in them, and easy to make mistakes. There are many difficulties and subtleties, and while some of these difficulties may be inseparable from the required calculation, others may be only present due to the baroque nature of the Standard Model.

It is helpful to distinguish unavoidable complexities from avoidable ones. The same problematic question may arise both in a theory with two particles and also in a theory with thirty-seven particles, and it is perhaps better and easier to understand it first in the former theory rather than the latter. Simplified toy models have a honoured place in physics, and one common approach in quantum field theory research has always been to understand its hard questions within controlled sandbox environments where calculations are tractable.

8.1   THE PROBLEM OF STRONG COUPLING

Of all the confusing but fun subtleties of quantum field theory, I will concentrate here on one ubiquitous and hard problem. This is the problem of understanding quantum field theories at *strong coupling*. What does this mean?

Many quantum field theories – for example, the Standard Model – are used to describe interactions between particles. If you fire beams of electrons at beams of positrons, quantum field theory tells you the ways these particles can interact and the probability of the different interactions occurring. These calculations are most easily done when all interactions between particles are extremely weak, and so there is little to no chance of double interactions. In chapter 3 we described this by using the metaphor of Arabella and Bert walking towards each other on the street. The case of very weak interaction is the case where Arabella and Bert are almost total strangers to one another – almost certainly they will carry on past one another to their respective destinations, and the chance of something happening is small.

The measure of interaction strength between particles is called a *coupling constant*.[1] For the quantum theory of electromagnetism, this constant is called the *fine structure constant* and has a numerical value of 0.0073 – approximately $\frac{1}{137}$. Different values for these coupling constants correspond to different quantum field theories. One can visualise these as being on a sliding scale, where we are free to slide the value of the constant up and down.

At one end of the scale is the limit of no interactions at all. This theory is called a free theory. It is the easiest limit of all. In this limit particles have no potential for interaction and pass each other as ghosts. In our analogy, Arabella and Bert are total strangers and walk straight past each other. As the constant is increased, we come to the point of weak interactions. There is a non-zero but small chance that particles will interact – but most of the time they still just carry on. Here Arabella and Bert are distant acquaintances, who nod, say hello, and go on their way. For this case of weak interactions, there is a stack of well-sharpened tools that can be used to compute the scattering probabilities. These tools are based on successive approximations: first the probability of no interaction at all, then for a single interaction, then two interactions, then three, and so on.

This procedure of successive approximations works well when the likelihood of any interaction is small. The higher interactions refine the answers, but do not change them qualitatively. The greater the required accuracy, the more interactions need to be included – but these only matter far to the right

---

[1] The use of the word 'constant' here is common but unfortunate language, as the value of these 'constants' depends, in a calculable way, on the energies of the particles that are interacting.

of the decimal point. The view obtained by including only the first term is fuzzier, but does not mislead.

For the case of *strong coupling*, this approach breaks down. The stratified hierarchy of calculational significance disappears. It is no longer possible just to compute the first terms of an approximation to obtain a rough picture of the final answer. Every term matters as much as any other – the case of sixteen interactions is on an equal footing with the case of a single interaction. To obtain the correct answer, we now need to find all the terms – an infinite number – and sum them up. Arabella and Bert are first cousins who have not seen each other for years: who can predict what will happen when they bump into one another?

It is important to say that being at strong coupling does not signal any intrinsic fault with the theory, which still makes perfect sense. The value of a parameter has been changed, but the theory is as well defined as it ever was. The problems lie with us, and our desire to calculate in the theory. The old tools have gone, and it is totally unclear what should replace them.

How to proceed? How to calculate? This problem of understanding quantum field theories at strong coupling is an old one that extends back fifty years. Its original motivation was the historic purpose of physics: to understand the natural world. For almost all purposes, one of the four fundamental forces – unsurprisingly, the strong force! – is strongly coupled. At any energy that was accessible prior to the start of the 1970s, the strong force lies in a regime where it is strongly coupled. At the time, this caused enormous bafflement. No one knew the defining equations of the strong force, or even had a clue where to start looking for them.

Ultimately, the crucial simplifying feature of the strong force was the fact that it is only strongly coupled at 'low' energies. As colliders became larger, the energies of the colliding particles increased and, *mirabile dictu*, the strong force ceased to be strong. It became weak. It became calculable. It became accessible. All that had been needed was an increase of the kinetic energy of the colliding particles to a level much greater than the rest mass energy of protons and neutrons – and then suddenly the strong force could be analysed with familiar tools.

This does not make it any easier to understand the strong force at low energies, but it does give some clarity to the question. The fundamental equations for the strong force do not have to involve strong coupling. There are energies at which the strong force is strongly coupled, and there are energies at which the strong force is weakly coupled. While it remains difficult to compute in the strong coupling regime, one can also in a certain sense quarantine these regions and hive them off as a 'bad' part of the theory where calculations are difficult. The hard problem remains hard, but its boundaries are now clearly marked.

Once upon a time, the motivation to study quantum field theory at strong coupling was the desire to understand the behaviour of the strong nuclear force at low energies. This historic rationale is now less pressing, and this particular task is no longer considered an important frontier problem in theoretical physics. The reasons for this are several.

First, the strong force really does describe nature. As a result, vast computing power has been thrown at it. By formulating the problem carefully and throwing enough processors at it, many aspects of the behaviour of the strong force at strong coupling can be found. This is not ideal. It cannot address every aspect, it is not always elegant and it is not always insightful: but it does give correct answers. Computers calculate rapidly, do not get bored and do not need sleep.

The second reason is that, again given the importance of the strong force, every possible cheap trick that can be used on it has been exploited. For example, the masses of the up and down quarks are far lighter than the masses of all other quarks. This did not have to be true and we do not know why it is true. It is a particular, unexplained feature of the quark masses in the Standard Model. However, it does lead to a series of calculational simplifications that apply for the actual strong force but cannot be extended to more general examples of strongly coupled quantum field theories. Whatever the restrictions on their broader use, these little dodges provide information about the actual strong nuclear force that arises in nature.

The final reason is that physics moves on. Not all problems are equally important. Effort requires motivation, and attaining funding requires a good answer to the 'Why are you working on this?' question posed by both fellow scientists and research councils. When there was *no idea* what the correct theory of the strong nuclear force was, it was unclear what the angle of attack should be – any direction may have produced the key that turns the lock. However, once the correct theory is established, it is not so important to compute everything that could ever possibly be computed. Complicated problems that require time and effort to solve require a justification for the investment of time and resources. The full gravitational dynamics of the solar system, including all the minute little wobbles from the other planets and their moons, is studied today to investigate its stability and to ensure that rockets reach their destinations, not because we want to test Newtonian mechanics.

Today, the chief motivation for understanding strongly coupled quantum field theory is not primarily to discover new laws of physics. New physics is generally discovered first through weakly coupled effects rather than through strongly coupled effects. This was true for all three forces of the Standard Model and one may wonder why. It is not that the complications of strongly interacting physics are never relevant for nature. It is more that the first signs of new strongly coupled physics can be described as new weakly coupled physics. New physics almost always appears as small deviations from existing known laws – and small deviations can be described through weak coupling approximations. The reason for this is as follows. If a deviation is large, under less sensitive instruments the deviation would still be present, but small. Given incremental progress in technology, any new physics will appear first as simply a small deviation from existing laws, and such small deviations can always be parameterised through weak coupling techniques.

As an analogy, we could ask what sound is made by a stampeding herd of rhinoceroses. The sensory experience at the centre of the charge of the pachyderms would indeed be striking. However, the first answer would always be as a soft, distant patter on the ground. The soft patter will turn to a quiet rumble, and the quiet rumble will turn to full thunder: but the patter will come first.

However it arose historically, the upshot is that there are a large body of physicists whose research is aimed at understanding quantum field theory at strong coupling, and who are interested in any tools that may be used to calculate in this region.

Based on what I have said above, it may seem hard to conceive how any calculations can be performed in this area, even in principle. This book is not a textbook and is almost free of equations. Nonetheless, I wish to try and illustrate how this can work, by describing one method of computation that can be, and is, used to give results in this regime. It is certainly not the only method, but it is a real method. I describe it to illustrate how, despite all difficulties, controlled computations can actually be performed within a strongly coupled theory.

This method is based on the fact that the strong coupling regime is obtained by sliding the value of a coupling constant from weak values to strong values. We suppose the theory starts in the weakly coupled regime, within which it is 'easy' to calculate. As the coupling is smoothly varied, the theory continuously moves into the incalculable strong coupling regime.

The key word here is 'continuously'. Whatever the change may be, it is continuous, and so whatever quantities we compute must also vary continuously as we go from weak to strong coupling. In particular, jumps are forbidden. However, suppose we consider a quantity that can both only take integer values – nought, one, two, three ... – and is also forbidden to jump. Such a quantity must take the same value irrespective of whether it is evaluated at either weak or strong coupling, since the change has to be continuous and the quantity cannot jump. If this quantity is computed in a theory with weak or even no interactions, it will have precisely the same value as in the strongly interacting theory. If the quantity is chosen carefully, this will give interesting information about the behaviour of the strongly interacting theory.[2]

Many people find strongly coupled quantum field theory interesting. However, none of what has been said so far has anything to do with either string theory or quantum gravity. The merry community of quantum field theorists could have happily continued on their merry way, innocent of gravity in two, four, ten or twenty-six dimensions. However they did not, and this merry

---

[2]One of the best-known examples of such an interesting quantity is called the Witten index, after Edward Witten. Roughly, it counts the difference in the number of bosonic and fermionic states in a quantum field theory.

community is now infested with ideas from string theory. This is because it turned out that certain problems in strongly coupled quantum field theory could be best attacked, and indeed solved, using higher-dimensional theories of gravity. Even if your sole concern in life was an understanding of quantum field theory, it became apparent that one of the most powerful tools was string theory – and as the millennium approached, even the most staid of quantum field theorists became drunk on the new wine of string theory.

8.2   THE ADS/CFT CORRESPONDENCE

The date of this wine's bottling was Thursday the 27th of November 1997 – Thanksgiving Day. On that day the Argentine physicist Juan Maldacena, then at Harvard, uploaded to the electronic preprint archive a single, short paper with the unprepossessing title *'The Large N Limit of Superconformal Field Theories and Gravity'*. Maldacena is a quiet and softly spoken man not given to large demonstrative gestures. The paper is technical, and on first glance might appear to be dealing only with minutiae. However on careful reading one realises that the paper makes a bold claim: certain quantum field theories are one and the same as certain gravitational theories, specifically certain string theories. 'One and the same' means exactly that – they are completely identical objects written in different notations. Any calculation in one can also be done in the other. In the language of chapter 3, there is a duality between the two formulations – gravitational theories can be dual to quantum field theories.

One way to quantify the impact of Maldacena's paper is through numbers. Since it first appeared, it has been cited by more than ten thousand other papers. This is a lot. To put this figure in perspective, every year around six thousand papers are published within theoretical high-energy physics. These figures imply that, over the eighteen years since it appeared, Maldacena's paper has been cited by around one in every ten of the hundred thousand or so papers that have appeared. Another, maybe slightly embarrassing, perspective on this is the fact that this adds up to more citations than any other paper in high-energy physics *ever* – including all the foundational papers of the Standard Model, such as Steven Weinberg's 1967 classic *'A Model of Leptons'* with 'only' nine thousand citations.[3]

A more graphic illustration of the pentecostal impact of this paper is through dance. Physicists do not, *pace* Feynman, have a reputation as party animals. In most people's estimation, an assembly of theoretical physicists would rank slightly below a gathering of IT database experts in the ordering of places to let your hair down. And yet at the Strings 1998 conference at Santa Barbara in California, the after-dinner speech at the conference banquet saw

---

[3] It is true that authors in the past cited less frequently and more discriminately. Inflation has applied to citation rates as well as the price of bread. While Weinberg's paper has fewer absolute citations, its inflation-adjusted number would be higher.

four hundred physicists dancing in unison as the speaker Jeff Harvey from the University of Chicago orchestrated a twist on the hit pop song 'Macarena':

> You start with the brane
> and the brane is BPS.
> Then you go near the brane
> and the space is AdS.
> Who knows what it means
> I don't, I confess.
> Ehhh! Maldacena!

> Super Yang-Mills
> with very large N.
> Gravity on a sphere
> flux without end.
> Who says they're the same
> holographic he contends.
> Ehhh! Maldacena!

> Black holes used to be
> a great mystery.
> Now we use D-brane
> to compute D-entropy.
> And when D-brane is hot
> D-free energy.
> Ehhh! Maldacena!

> M-theory is finished
> Juan has great repute.
> The black hole we have mastered
> QCD we can compute.
> Too bad the glueball spectrum
> is still in some dispute.
> Ehhh! Maldacena!

It was not quite the chairman of the Federal Reserve doing the Can-Can in stockings and suspenders, but it has never been seen again.

What precisely is this correspondence? As said above, it is the claim that certain gauge theories – quantum field theories of the kind that are used to describe the strong, weak and electromagnetic forces – are exactly the same theory as certain gravitational theories, just written in a different language.

The equations used to write down the two theories are totally different. The types of interactions that are present are totally different. Even the number of dimensions in the two theories are different. In this correspondence, it is always the case that the gravitational theory has one extra dimension than the quantum field theory it is equivalent to. The claim that the theories are exactly the same is therefore shocking and audacious. Although I have just used the word 'exactly' twice already, it is good to repeat it a third time. This is not a statement that one theory is a metaphor to describe the other, or that sectors of one theory provide a useful approximation to the other. It is a statement that the two theories are completely and utterly equivalent. They are exactly the same theory, simply re-expressed in different terms.

In Maldacena's original paper, this correspondence was first applied to one particularly special quantum field theory, called four-dimensional super Yang-Mills theory. This theory does not itself describe the real world, but is a generalisation of the theories that do. It is a souped-up version of the theories that describe the strong, weak and electromagnetic interactions, and what makes it special is the large amount of symmetry it has. It has both the maximal amount of symmetry and the maximal amount of supersymmetry possible.[4] It is to the Standard Model what a flawless cut diamond is to rock pulled fresh from a mine. The enormous amounts of symmetry highly constrain its structure and allowed interactions, and it had already been singled out as a 'special' quantum field theory worthy of particular study even when Juan Maldacena was a young boy in short trousers in Buenos Aires trying to understand how the radio worked and helping his father fix the family car.

This special theory also comes with two parameters. The first is an integer imaginatively denoted by N, which relates to the number of 'force carriers'. In electromagnetism there is one force carrier, the photon. In the weak force there are three, the W^+, W^- and Z^0 bosons. For the strong force, there are eight force carriers – the different kinds of gluon. In our special theory of interest, a value of N implies $N^2 - 1$ force carriers. The second parameter is a continuous coupling constant, which determines the strength of the interactions.

In its simplest form, the correspondence says that this special theory – maximally supersymmetric four-dimensional Yang-Mills theory – is exactly the same as one of the string theories, type IIB, living on a ten-dimensional space in which five dimensions make up a sphere and the remaining five dimensions form a particular geometry called Anti-de Sitter space.[5] The presence of two parameters, one continuous and one an integer, is mirrored in the string theory. The continuous parameter this time is the string coupling constant, which describe the likelihood that two strings will interact with each other. The integer parameter also exists in the string theory – but where it comes from and what is means is far too technical to discuss here.

These last few paragraphs have been dense. If they have been too dense, it is important to return to the basic claim: a quantum field theory in four dimensions with no gravitational interactions is entirely equivalent to a gravi-

---

[4]It is not possible to add any more supersymmetry to the theory without turning it into a gravitational theory. It therefore sits in a unique position as the theory with the largest possible amount of supersymmetry – but which is not yet a gravitational theory.

[5]'Anti' here does not have the same connotations as in anti-particle: there is no sense in which de Sitter space and Anti-de Sitter space can annihilate with each other!

tational string theory in five dimensions.[6] The claim is that any result present in one theory can also be obtained in the other, once re-expressed in the right language.

Having the same information expressed in different languages is of course familiar to scholars of the humanities, and in certain cases exceptionally useful – as for the Rosetta stone, where the presence of the same text in Demotic, Ancient Greek and Egyptian hieroglyphics enabled the last to be deciphered in the early 19th century and thereby read for the first time in fifteen hundred years. Both linguists and scientists need a dictionary to translate from one language to another. In the physics case of the AdS/CFT correspondence, this requires an ability to say 'Quantity A in the gravity theory corresponds to quantity B in the gauge theory. To calculate quantity A in the gravity theory, it is sufficient to calculate quantity B in the gauge theory, for they are one and the same.'

If Maldacena's paper served the role of the Rosetta stone, two papers in February 1998, one by Steven Gubser, Igor Klebanov and Alexander Polyakov, and another by Edward Witten, all at Princeton, started to fill in the entries of the dictionary. While Maldacena's paper had claimed the equivalence of the two theories, these papers gave calculational prescriptions for checking this equivalence, with statements of precisely which calculation in the gravity theory was equivalent to precisely which calculation in quantum field theory. A further paper by Witten the following month filled in some more of the dictionary. In particular – and this will be important later – he filled in the entry for 'black hole'. A black hole in the gravity theory, Witten said, is equivalent to heating up the quantum field theory so that it is at a finite temperature.

In many cases, the prospect of curling up in an armchair to read a dictionary is less than enthralling. What made this dictionary a scientific bestseller was the fact that it identified strongly coupled quantum field theory with a weakly coupled gravitational theory.[7]

This makes the duality not just of conceptual interest, but also useful. The duality identifies strongly coupled quantum field theories, in which it is hard to compute, with classical gravitational theories, in which it is easy to compute. This allows the hard problem of computing in quantum field theory at strong coupling to be transmuted into the 'easy' problem of computing in classical general relativity at weak coupling.

---

[6]The statement 'string theory in five dimensions' is slightly loose: as alluded to above, is better to say that the string theory is in ten dimensions, and five of those dimensions are curled up into a sphere while the other five dimensions form Anti-de Sitter space.

[7]There is a technical subtlety here. The quantum field theory actually has two couplings, one related to the interaction strength and one related to the number of particles. In professional language, these are the gauge coupling and the 't Hooft coupling. Likewise, the gravitational theory also has two couplings, one relating to the interaction strength of strings, and one to the size of strings relative to the ambient geometry. In the canonical example of the correspondence, both couplings in the gravitational theory are small.

It is this fact that is of most relevance to this chapter. Even if you cared not a whit about gravity, even if you are cold to any and all ideas about extra dimensions or black holes or mathematics: if you are interested in calculating in general quantum field theories, you should be interested in string theory. This was the carrot that attracted many workers in quantum field theory to learn about string theory and spend time on it, all for the furtherance of their own interests.

All very important – if correct. But how do we know, and why should we believe, that the correspondence is correct? There are many, many ideas in science whose path to importance and influence is blocked only by their falsehood. What makes AdS/CFT different?

This is an important question which deserves an answer, to be given over the next few pages. The first point of clarity is that it will not be a philosophical answer. The question of how ideas become accepted, and how we can acquire justified true belief, is an interesting question within philosophy and epistemology – but the issues raised there apply equally well to the statements 'The sun will rise tomorrow' and 'I am the son of Tom and Theresa Conlon'.[8]

The second point of clarity is that it will also not be a mathematical answer. Mathematicians work with precise and well-defined objects. The AdS/CFT correspondence involves on the one hand string theory and on the other hand quantum field theory. Neither side of the correspondence involves an object with a rigorous mathematical definition. Forty years after the correct theory of the strong force was established, there remains a one million dollar prize from the Clay Mathematics Foundation for showing that this theory exists mathematically and determining some of its (well-established) properties. The mere existence of this prize illustrates the difficulties in giving a mathematical definition of a theory as well established physically as quantum field theory. The difficulties for string theory are proportionately harder; there is not even a starting point for writing down rigorously a mathematical object that represents 'type IIB string theory on the product of a five-dimensional sphere and five-dimensional Anti-de Sitter space'.

A mathematical proof of the correspondence must involve a construction of mathematical structures representing both sides of the correspondence – one structure for quantum field theory and one structure for string theory – and then a formal proof that these two structures are the same, or in the language of mathematics, isomorphic. As no one knows how to construct either mathematical structure, no such mathematical proof exists, or indeed will appear anytime soon.

---

[8]There exists all kinds of philosophical agonising about the scientific process and the inductive method. One example: define 'grue' as the colour 'green until today; blue from tomorrow'. The statements 'grass is grue' and 'grass is green' have identical empirical support, although we intuitively think the latter correct and the former unscientific. The jury may be out on whether this represents an intellectually productive train of thought, but for right or wrong physicists note these concerns in the way the Levite on the road to Jericho noted the naked and robbed traveller.

The answer I will give is instead that of a physicist. It is that you check the correspondence by calculation, you carry on checking the correspondence by calculation, and you continue checking by looking for as many ways as possible of breaking the correspondence. At some point – and this point will vary from person to person – sufficiently many checks have been performed that it becomes hard to think of any way the correspondence could both pass all these tests and still not yet be true. This is not proof. No number of checks ever make a proof. However physics is not mathematics, and those with scruples on this matter can be well advised that the math department on campus is generally in the next building down the street.

The tests build up in complexity. The simplest version of the correspondence involves on one side an entirely classical gravity theory called ten-dimensional supergravity, and on the other side the maximally special Yang-Mills field theory in the limit of an infinite number of particles. The calculational content of the correspondence relates on the gravity side masses of particles in the gravitational theory, and on the field theory side the way certain quantities change on moving from large to small distances. This last expression sounds somewhat vague, but it does have a precise technical definition as what are called anomalous dimensions of operators. The check in this limit has many moving parts, and it is by no means trivial – but it works.

These simpler tests can be extended. The gravitational theory moves away from being merely a supergravity theory, with equations coming from a glitzed-up version of Einstein's theory of general relativity, to being a full string theory. The simplest version of string theory is a classical string theory – an extension from general relativity to string theory, but where all the quantum effects have been switched off. The correspondence works. The tests can then be extended again to quantum string theory – and again the correspondence works.

These progressively more complicated tests give progressively more impressive evidence that the AdS/CFT correspondence really is true, and that the two objects it claims to be identical are indeed identical. Of course, it is true that these calculations are performed using a standard of rigour common in theoretical physics rather than in mathematics. It is also true, as said before, that the accumulation of any number of successful tests of the correspondence is mere inductive evidence, offering no logical proof of correctness – as with the examples of 'grue' and 'green' in the previous footnote.

Everything I have just said is words. However I do want to try and illustrate just *why* physicists find these tests so compelling, by moving beyond words to include an equation. I do so in the hope of providing a sniff of what is so powerful about this correspondence, by showing just how non-trivial this agreement is. I will write below an equation representing the result of a calculational tour-de-force: a five-loop calculation in super Yang-Mills theory. To professionals in the field, 'five loop' is to calculational difficulty what the Strongest Man in the World competition is to shifting some

Christmas flab from your tummy. No higher level of precision has been attained for quantum electrodynamics, the oldest and simplest element of the Standard Model. The equation is[9]

$$\gamma_K(a) = 3a - 3a^2 + \frac{21}{4}a^3 - \left(\frac{39}{4} - \frac{9}{4}\zeta(3) + \frac{45}{8}\zeta(5)\right)a^4$$

$$+ \left(\frac{237}{16} + \frac{27}{4}\zeta(3) - \frac{81}{16}\zeta(3)^2 - \frac{135}{16}\zeta(5) + \frac{945}{32}\zeta(7)\right)a^5.$$

Here 'ζ' refers to a particular mathematical function called the Riemann zeta function, named after the 19th century German mathematician Bernhard Riemann, and $\zeta(3)$, $\zeta(5)$, and $\zeta(7)$ are the particular values this function takes for values of 3, 5, and 7.

What matters here is not so much what the equation means as the rather intricate form it has. Look how complicated it is! This equation has been computed in two different ways, once using the techniques of quantum field theory and once using the techniques of string theory – obtaining perfect agreement. This is almost immediate evidence that something deep is happening – it is not possible to write a term like $\frac{945}{32}\zeta(7)$ by chance, and it is clear the agreement between these calculations is no mere accident.

This is not a solitary example. There are many similar cases where the same quantity has been computed in two different fashions and in two very different theories, both on the field theory side and on the gravitational side, and complete agreement has been attained. As with this example, what is psychologically convincing is the deeply non-obvious form of the expressions encountered. These are not factors of two or minus signs. The expressions are the product of arduous calculation and involve results one could never guess in advance. It is striking that two different calculations, done in different theories and using different methods, give the same answer. This test of veracity is the theoretical equivalent of when two experiments on two continents use two techniques to measure the same physical quantity and get the same answer. When this happens many times, it does not prove a thing – but it does convince.

In fact, life is even better than this. The special theory referred to above is indeed special. Many of its properties have been studied extensively. This means that in some cases it is possible to obtain exact results for this theory. This is not possible in general quantum field theories and is only possible here because this theory has many more symmetries than a normal quantum field theory. Symmetries constrain. They enforce relations between quantities with no other connection. The more symmetries, the more constraints. For this particular theory, the symmetries can sometimes be enough to constrain everything so well that in the end there can only be one result: the exact answer. An exact answer in quantum field theory is normally an unattainable dream,

---

[9]For experts: this is the anomalous dimension of the Konishi operator in the limit of an infinite number of colours.

and this is why all the many approximation schemes have been developed. However, the specialness of this special theory does in some cases allow for exact solutions, to which the five-loop quantity just described is 'merely' a leading approximation. For such exact quantities, it is possible to interpolate results smoothly between the regions where the gravity description is 'easy' and where the quantum field theory description is 'easy'.

This section has described checks of the correspondence for its canonical avatar, between the maximally symmetric quantum field theory in four dimensions and a particular string theory in five dimensions. However, while this case offers the most precise arena for a critical test, there are many other examples as well. These include cases of quantum field theories in two or three dimensions that are equivalent to gravitational theories in three or four dimensions. It also includes many cases of 'less symmetric' quantum field theories in four dimensions. These theories have more in common with a theory like the Standard Model but also have extra symmetries that constrain the results and make calculations more tractable. Again, the correspondence has been tested, and has not been found wanting.

To recap: while there is no formal proof of the AdS/CFT correspondence, most physicists are happy with its correctness. This is because it works. There are solid physical arguments for why it should work, and its implications have been validated in numerous disparate environments. The tests are not trivial, and involve the reproduction of intricate formulae with no obvious structure. At some point, and after many many successful tests, the question shifts from 'Is there any interesting way this idea could possibly be correct?' to 'Is there any interesting way this idea could possibly be wrong?'.

Given it is correct, what should we use it for? One application is to understand quantum field theory for the sake of understanding quantum field theory. While this is a worthy goal with many papers written on the topic, it would be nice to apply the AdS/CFT correspondence to real systems in the real world. This is the focus of the next two sections.

8.3   APPLICATIONS TO COLLISIONS OF HEAVY IONS

According to legend, Hermes Trismegistus managed it. Isaac Newton tried – and failed, as did his contemporary Robert Boyle. What is 'it'? 'It' is the fabled goal of alchemy – the ability to turn base metals into gold.

Newton and Boyle did not fail in their alchemical experiments because there was something intrinsically stupid about their attempts. It is only heat and pressure that divides pencil shavings from diamonds, and both cooks and cement manufacturers know well how easy it is to change the texture, colour and hardness of an object. It is true that the alchemical quest is one marked by many failures, and it is true that the sane and reasonable give up after decades of trying and not succeeding. However it is also true that the sane and reasonable are often not the ones who make major discoveries.

Alchemy fails for reasons that were not known, and could not have been known, at the time of Newton. It fails because the transmutation of base metals to gold is not a chemical change but a nuclear change. In a chemical change, the bonds between different atoms and molecules are reordered and rearranged. In a nuclear change, it is the bonds between the protons and neutrons that make up nuclei that are broken and forged anew. It is an empirical fact that chemical changes need energy inputs that can be obtained from domestic sources of temperature and pressure, while nuclear changes require inputs of energy going far beyond those domestically accessible.

Given the amount of scientific effort that went into attempts to make gold, it is somewhat ironic that experiments now exist that are dedicated to the destruction of gold and the dispatch of its constituents whence they came. These experiments take atoms of gold and strip them of all their electrons. This leaves a large denuded gold nucleus of around two hundred protons and neutrons, carrying seventy-nine units of positive charge. With so much charge, the nucleus is easily accelerated within a magnetic field. Indeed, these gold nuclei can be accelerated by the same magnets that are used to accelerate protons, which hold one unit of positive charge, and in the same accelerators – such as the Large Hadron Collider.[10] In precisely the same way that the Large Hadron Collider accelerates two counter-rotating beams of protons and collides them into each other, it can also accelerate and then collide two counter-rotating beams of nuclei. The effect of the collision is exactly what you would expect: the two nuclei and their constituents are utterly destroyed. It is a nuclear Götterdammerung of the highest order. Each gold nucleus starts with a total of around two hundred nucleons – and in the collision they are entirely destroyed and broken into their constituent quarks and gluons.

What happens next? And why do it in the first place? The reason for performing these collisions is not to enhance through scarcity the remaining value of the world's gold supply: even at continuous operation it would take longer than the age of the universe to do so. Neither is it to satiate an infantile urge of destruction for the sake of destruction. The purpose of these collisions is to create a state of matter that last occurred naturally only a fraction of a second after the Big Bang. This state of matter is called the *quark-gluon plasma*, and it consists of a hot thermal bath of quarks and gluons. This bath can be formed only in temperatures that are hot enough to melt nuclei and dissolve the strong force bonds that normally bind the quarks and gluons.

In this respect, the formation of the quark-gluon plasma resembles the formation of an ordinary bath of water starting from a tubful of ice cubes. If the bath is placed outside on a summer's day, the ambient temperature is large enough to break the chemical bonds that hold the water molecules together as ice, and produce water. The quark-gluon plasma is formed in a similar fashion, except it is nuclear bonds that are melted by high temperatures, and

---

[10]Strictly, the Large Hadron Collider uses lead ions while its American counterpart the Relativistic Heavy-Ion Collider (RHIC) at Brookhaven National Laboratory uses gold.

the temperatures required to sustain this state are now a hundred thousand times greater than those present at the centre of the sun.

Once upon a time in the early universe, these temperatures were everywhere. The quark-gluon plasma remained in existence during this period, as there was no need to worry about a hot patch being cooled by contact with a cold patch. It is never fully accurate to say that the Large Hadron Collider recreates the Big Bang, but this is the least inaccurate sense of the statement. Through forming the quark-gluon plasma, the Large Hadron Collider replicates for a brief instant the conditions that held in the very early universe.

That was then. As described in chapter 2, as the universe grew it cooled, and in the here and now such temperatures can no longer be sustained.

However, just for a brief moment, one can imagine it is the early universe again. When two high-energy gold nuclei are smashed into each other, for a short period after the collision all its energy goes into the quarks and gluons that are liberated. It heats them up. Quark hits quark that hits gluon that produces anti-quark that hits quark that makes gluon that hits anti-quark. All the produced particles collide rapidly with each other and thermalise. There are around four hundred nucleons initially, and immediately after the collision there may be many thousands of particles present. The concept of temperature requires many particles. Ten thousand is not a million billion billion, but it is morally far closer to it than it is to two or three. Indeed, ten thousand is certainly enough to briefly form a little region of hot quark-gluon plasma.

This little region does not survive very long. Like high-pressure superheated steam emerging into an Arctic winter, it rapidly expands, dilutes and cools down. As it cools, the quarks and gluons recombine to form protons, neutrons, pions, kaons and anything-you-likeons. However, deep inside the accelerator, for a twinkling the quark-gluon plasma was re-made, and its empirical properties can then be inferred from the debris recorded by the surrounding detectors.

The purpose of these experiments is to study the quark-gluon plasma. Two such experiments exist. There is one dedicated experiment, the Relativistic Heavy Ion Collider (RHIC) at Brookhaven National Laboratory, and one part-timer, the ALICE experiment at CERN's Large Hadron Collider. ALICE is the grubby night-shifter of the four big CERN experiments, allowed to sneak one month of heavy ion data a year away from the glitterati of the two general purpose experiments ATLAS and CMS.[11] And the purposes of studying the quark-gluon plasma? These are twofold. The first reason was just described: the quark-gluon plasma is a phase that the universe passed through very early in its history. By re-creating this phase, we can study its

---

[11]The acronyms used in particle physics are not imaginative. ATLAS is A Toroidal LHC AppraratuS, CMS is a Compact Muon Solenoid and ALICE is A Large Ion Collider Experiment.

properties experimentally and thereby better understand the physics of some of the earliest moments of our universe.

It is the second reason that fits into the story of this chapter. This is the fact that the quark-gluon plasma is another phase of the strong force, quantum chromodynamics, in the same way that ice and steam are other phases of water. For those who are interested in the quantum field theory of the strong force, the quark-gluon plasma gives them a chance to study it experimentally in a novel setting. The quark-gluon plasma fills in part of the map of quantum field theory and part of the map of the strong force, making it interesting to those who care about understanding these matters.

These experiments are then interesting to the community of people who want to understand what happens in collisions of heavy nuclei at high energies, and also to the overlapping but not identical community who want to understand quantum field theory at strong coupling and finite temperature. String theory has become interesting to them because it provides a novel perspective on the quark-gluon plasma that has had some calculational success in describing real-world experimental phenomena.

String theory does so by providing, through the AdS/CFT correspondence, a dual description of the hot mess that is produced in the collisions of heavy nuclei. The dual description provides novel calculational techniques for the hot thermalised fireball of thousands of quarks and gluons that make up the quark-gluon plasma.

As we have seen, the basic mantra of AdS/CFT is that the dual of a strongly coupled field theories is a weakly coupled gravity theory. One may wonder what corresponds to temperature – what object is the gravitational dual of a *hot* field theory? We mentioned the (initially surprising) answer a few pages ago. The answer is a black hole. The fact that the field theory is hot, and at a constant temperature everywhere, is dual to the statement that a black hole is present within the gravitational solution. This sounds odd, but there are sanity checks that show that this is, if not obviously correct, not obviously wrong either.

First, both a black hole and a completely thermalised plasma are endpoints. For both, the gate of entry is tall and wide, and the gate of exit short and narrow. Suppose you start with one region of the plasma at slightly higher temperatures, another at slightly lower, and a third superheated bubble in the middle. The laws of thermodynamics tell you that this inexorably leads to the hotter regions transferring heat to the cooler ones, such that in the end all are at the same temperature. As with two taps feeding into a sink, the combination of an influx of hot and an influx of cold results in everywhere warm. Whatever the initial temperature variations are, given time you end up with a uniform temperature bath.

In a similar way, a black hole is the final destination for many gravitational problems. Matter attracts via gravity. The more matter you have, the more it attracts. As matter comes closer to other matter, the gravitational attraction becomes larger and larger. Eventually, when there is sufficiently much stuff in sufficiently small a region, it collapses to form a black hole. At this point all details of the original configuration disappear and what remains is a featureless endpoint with no trace of whether we started with dust or diamonds. Both a uniform thermal bath and a black hole therefore share the property of being an end rather than a beginning, an omega rather than an alpha.

There is a second reason to motivate why a black hole might sensibly be the dual description of a field theory at a finite temperature. This is the fact that black holes are not actually black. In classical physics, black holes only swallow. In quantum physics, they also spit. This shocking result is due to Stephen Hawking, who in addition showed that the spectrum of the light they emit is that from an oven at a fixed temperature: the Hawking temperature. The 'temperature' property of the field theory is mirrored by the 'temperature' property of the black hole. Small black holes have big temperatures and big black holes have small temperatures.

None of this makes a proof. However, it may provide more motivation for the idea that, in the AdS/CFT correspondence, the dual to a field theory at a finite temperature is a gravitational system with a black hole. This is conceptually interesting – but so what? What gives this relationship teeth? The teeth come from asking the question: what interesting properties can be computed about black holes? What do these properties correspond to on the field theory side?

One happy benefit of the duality is that black holes come with a totally different set of intuitions as to what is easy, what is hard, and what 'must' happen than do lumps of hot plasma. Indeed, black holes are rather universal objects. Unlike fingerprints, faces and DNA, it is not the case that every black hole is unique and with a large number of distinguishing features. While humans may be unique, black holes are identical up to a relatively small number of parameters: their mass, their rate of spin and their electric charge. This is a famous result of black hole physics, laconically expressed as 'black holes have no hair' and known as the no-hair theorem. It is as if all human faces could be labelled by age, skin colour and eye colour – and these alone and nothing else. This universality result implies that, for many properties, if you calculate for one black hole, you calculate for all black holes. There is no need for a separate calculation for each new black hole.

Why is this interesting? The AdS/CFT correspondence is best defined for theories that have no precise realisation in the real world. These are theories with an infinite number of particles and special symmetries that restrain the form of allowed interactions. These theories are related to the real-world strong force that appears in the quark-gluon plasma, but only as second or third cousins. The advantage of computing 'universal' quantities is that it offers a hope of finding the quantities least sensitive to the difference between the theory you want to calculate in and the theory you can calculate in. All our results about black holes are obtained for classical Einstein gravity. This is not dual to the real strong force. However, the hope is that perhaps the difference is small enough that one can use Einstein gravity as an approximation to the real theory.

What universal quantity is most interesting? The answer to this question involves a story that starts back in 1985 within the student dormitories of Moscow State University. Moscow was the political, cultural and intellectual centre of the Soviet Union. From all over both the USSR and the allied socialist countries, the best and brightest of young high schoolers came to study at Moscow State to be taught physics the Russian way.[12] Two of these were Dam Thanh Son from Vietnam and Andrei Olegovich Starinets. The students were grouped in dorm rooms by subject, and Son and Starinets were two of four physicists sharing a room. The assumption was that the Soviet students would assist their foreign colleagues to get better acquainted with the country and the language. Son was quiet and unassuming, but soon revealed his intellectual mettle. At one point in the first week, he was sitting quietly in his corner, reading a Russian-Vietnamese dictionary while the other students discussed a problem in complex analysis. He gently intervened, spent a few seconds explaining the solution, and then retreated back to his corner.

Son and Starinets remained roommates for the next three years. As they advanced in their studies they began to go their separate ways. Son became a graduate student at the Institute for Nuclear Research, and he and Starinets would now only occasionally meet in the dining hall. Then came the dissolution of the Soviet Union, and the collapse of the Soviet scientific culture. Many scientists emigrated to ensure a continued paycheck, as well as for other reasons, resulting in a diaspora of Soviet-trained scientists across the globe. Both Son and Starinets would end up in the United States, Son working on nuclear theory and Starinets working on string theory.

Many student friendships reduce with the passage of time. Moving forward to the turn of the millennium, Son and Starinets had not seen each other for many years. However they were now once again in the same city: Starinets was coming towards the end of a PhD at New York University, while Son had just started as an assistant professor at Columbia University several miles away. Starinets was working with his fellow student Giuseppe Policastro on the probability that gravitons – elementary quanta of the gravitational force – would be absorbed or scattered by black branes, string theory relatives of black holes. This represented a generalisation of earlier work by Igor Klebanov of Princeton University to the case of a finite temperature. By March 2001 they had finished the calculation: the question was, what could they do with it? They knew that under the AdS/CFT correspondence it said something about quantum field theory, but what was it?

---

[12]Russian theoretical physics developed a distinctive style that produced many first-rate scientists. Its spiritual leader was the scientific polymath Lev Landau, who got to live in interesting times, being born in 1908 and dying in 1968. The holy text of Russian physics was the ten-volume Course in Theoretical Physics by Landau and his student Evgeny Lifshitz. Russian physics is also famous for its seminars, which can last many noisy hours until every claim of the speaker is either satisfactorily settled or refuted.

Starinets suggested consulting Son, who he knew had already worked on finite-temperature quantum field theory. Policastro and Starinets headed up-town to Columbia on March 23rd, 2001 as one of their colleagues was also giving a seminar there. After the seminar, they found Son's office, knocked on the door, and told Son about the calculation. Son had been working on nuclear theory and had little prior knowledge of the AdS/CFT correspondence. The physicists talked together for several hours about the results and what they might mean. No definite conclusion was reached, but the ideas were buzzing. Policastro and Starinets returned home. That same evening, Starinets received a short email from his old friend:

Andrei,

I think I can relate $< F^2 F^2 >$ correlator with the viscosity.

Also, the bulk viscosity of the N=4 SYM is probably 0 because of scale invariance.

Is there a good review of AdS/CFT correspondence that you would recommend?

Son

Within a week the three physicists had a draft of a paper, which was then submitted to the preprint archive on 6th April, 2001. Why did they move so quickly and what were they excited about?

Son saw that the equations that Policastro and Starinets were using were not just calculating the absorption of gravitons in the gravity theory. They were also calculating something else: the shear viscosity of the field theory plasma it was dual to. For any kind of fluid, shear viscosity is a measure of how easy it is to slide the fluid past itself, or how easy it is for the fluid to flow. All fluids have some level of viscosity. If we treat water as a standard and familiar example, honey, tar and Marmite have high levels of viscosity, while air has a low level of viscosity. What was being computed here was the shear viscosity of the maximally supersymmetric field theory at high temperature.

The three physicists found that the viscosity of the field theory was proportional to the area of the black hole the field theory was dual to. All black holes have a horizon: the region of space which, once crossed, leads inevitably to a fall into the singularity at the centre of the black hole. The area of the black hole is simply the direct measure of how big the horizon is, and calculations show that the area of the horizon is set by the mass, or equivalently the temperature, of the black hole.

An important quantity associated to any black hole is its entropy. The entropy is, roughly, a measure of the total number of internal configurations that can possibly make up the black hole. The entropy was computed by Jakob Bekenstein and Stephen Hawking in 1973 and is also given by the area of the black hole. In the AdS/CFT dictionary, entropy on the gravity side goes across to entropy density on the field theory side, and so shear viscosity over entropy on the gravity side goes across to shear viscosity over entropy density on the field theory side. So, ultimately, what Policastro, Son and Starinets were calculating was the ratio of the shear viscosity of the field theory – its resistance to deformation – to its entropy density, roughly its level of disorder.

While these original calculations applied only to the special, maximally supersymmetric theory, it would subsequently be realised that the calculations applied not just there, but also for *any* quantum field theory whose dual gravitational description was a classical one involving classical black holes. Any field theory that was related by the AdS/CFT correspondence to such a gravity theory gave the same result. The result was applicable not just to one exceedingly special theory, but to an entire class – probably an entire infinite class – of field theories.

However, I would be fibbing if I said that these results were greeted by the world with rapturous acclaim. The result of Policastro, Son and Starinets was initially almost entirely ignored. For the eighteen months following their paper's appearance, not one single paper made any reference to it, except for those that had been written by the authors themselves. If anyone else cared, they were doing a good job of hiding it.

This paper now has almost one thousand citations. What changed, and why is this result now regarded as so interesting? To understand this, we return to what was calculated: a measure of the viscosity of something like the experimental quark-gluon plasma in theories that are, while not identical, related to the actual strong force. The calculation was also at large values of the coupling, in the precise region where it is extremely hard to calculate in the actual strong force. For reasons that are too technical for this book, this problem was also not amenable to the 'put it on a computer and press enter' approach. While there are aspects of the strong force at strong coupling that can be solved through cracking the whip and slave-driving a harnessed team of CPUs, this is not one of them.

While brute force could not be used, there had existed analytical approaches to computing this viscosity, which were based on using techniques applicable at weak coupling and extrapolating to strong coupling. Both the AdS/CFT approach and the weakly-coupled analytical approaches gave estimates for the shear viscosity of the quark-gluon plasma. These estimates were then tested by measurements at actual experiments such as the Relativistic Heavy Ion Collider at Brookhaven National Laboratory. The surprise was that the measurement of the viscosity disagreed badly with that computed in the 'traditional' approach – but agreed approximately with that computed using AdS/CFT. The viscosity of the quark-gluon plasma was surprisingly lower than expectations – and the only approach that seemed to get it right was that of AdS/CFT.

It is essentially this fact that drew so much attention to this work and so many physicists to this area. Physicists like to calculate. It is what they do. AdS/CFT offered an approximate model of the quark-gluon plasma that gave approximate agreement with various aspects of experimental data. The AdS/CFT model was amenable to calculation, and also to extension through the generation of further accessible variants. It offered a whole new set of insights into how to think about the strong force at finite temperature. As a bonus, it was connected to ideas about quantum gravity, but this was inessential and one could also take a purely pragmatic view. It was relevant to the quark-gluon plasma, and even if all one cared about was the quark-gluon plasma, that was enough.

Certainly it was not a perfect or precise approximation, but it also seemed better than the otherwise available options. There are many things 'wrong'. The theory the calculations were done in is not the actual strong force. The number of particles in this theory is far greater than in the actual strong force. There are additional types of particles that are not present for the real strong force: for example, scalar particles. The interactions are different. The symmetries are different. It is a model – but a model that has some agreement with the data.

However, this is precisely the *modus operandi* of physics: start with simplified and calculable models, and then expand through progressively adding refinements that make the models less and less simplified. This is how physics works. There is now a significant community working on using AdS/CFT as an approximate way to understand the strongly coupled quark/gluon plasma at high temperature. They use it both as a computational tool and for conceptual insight. Is it 'right'? It is useful, and that is what matters.

8.4   APPLICATIONS TO CONDENSED MATTER PHYSICS

The quark-gluon plasma belongs to particle physics, and quantum field theory is certainly a major part of particle physics. Quantum field theory has however spread its affections widely, and it is a subject with many lovers.

In particular, it is also pervasive throughout condensed matter physics. This is the branch of physics that describes phases of matter, both common and uncommon, that consist of lots and lots and lots of atoms. These include old favourites from school such as solids, liquids and gases, but also more exotic examples such as superconductors and topological insulators. These different phases have radically different properties and can behave in bizarre ways. What they have in common is that their meaningful existence requires many atoms to be present.

Confronted with a lone atom, it makes as much sense to regard it as a liquid atom or a solid atom as it does to view it as a rat atom or an elephant atom. By itself, it is just an atom. When many atoms are put together though, the behaviour of the collective exhibits distinctive properties. This is directly analogous to the movements of a crowd at a music festival or the flight of a murmuration of starlings. For this reason, condensed matter physics is also often called many-body physics. It is the science of how macroscopic properties of matter arise from the collective microphysics of large numbers of atoms.[13]

The number of bodies – atoms or molecules – involved in typical macroscopic systems is not just large. It is enormous. A glass of water contains enough atoms to put ten thousand of them on every square millimetre of the earth's surface. If we tried to write down the equations for each atom, the sun would be dead and cold before we were one millionth of the way through. The differences between solids and liquids, real as they are, cannot be found by direct assault on the equations of each individual atom.

This is an old problem, with a history that predates the advent of quantum mechanics. The motion of an ideal gas – separate individual atoms interacting with the rigid formality of billiard balls – was understood in Victorian times. This was a classical problem, but in many cases it is now necessary to consider the simultaneous effects of both quantum mechanics and the presence of many bodies. For close-packed atoms in a solid, with an average separation of somewhere around a nanometre, the distances are small enough that quantum mechanics cannot be ignored.

The study of quantum many-body systems has been carried out by many fabulously smart people over many decades. I will not and could not review it in any detail. Much of it is too tangential to this book, and my knowledge of the history is only of the mythological kind produced by textbooks and the names of Nobel Prize winners. I move straight to two important and relevant conclusions: first, quantum mechanics is essential, and second, quantum field theory is essential. The quantum field theories in this case are not the 'fundamental' quantum field theories that the Standard Model purports to be: these are practical, emergent theories that arise in the description of the physics of the many many atoms present.[14] In the Standard Model, displacements of the electric and magnetic fields from zero turn, through quantum mechanics, into particles of light: photons, the quanta of electromagnetism. In condensed matter physics, the displacements of atoms from their equilibrium positions in an atomic lattice also turn, through quantum mechanics, into particles: phonons.

Whether this physics is 'fundamental' or 'emergent', it is again a requirement to grok quantum field theory that sits athwart the road to understanding. This fact underlies a long history of intellectual interchange between particle

---

[13]Whether the resulting science should be called 'fundamental' is the subject of a long and mostly tedious debate. The 'No' argument is that many-body physics is only the study of the consequences of existing laws, and so does not by itself make new physical laws. The counter-argument is that the move from few to many is itself moving to a different regime of physics, in which the laws of few-body physics provide no guidance as to behaviour. The 'no' and 'yes' sides of this debate are associated with the words 'reductionism' and 'emergence'. A cogent and elegant argument for 'Yes' can be found in the article *'More is Different'* by Phil Anderson, one of the great names of condensed matter physics.

[14]As outlined in this book, the modern view is that the Standard Model is also in fact an emergent theory arising from a deeper underlying layer (such as string theory).

physics and condensed matter physics, starting as the centrality of quantum field theory to both subjects arose in the 1960s.[15] One of the key questions of that era was how quantum field theory behaved across different scales. As discussed in chapter 3, the conceptual framework for this was provided by the Nobel Prize winner Kenneth Wilson, who straddled both particle physics and condensed matter physics and solved important problems in both subjects. Wilson was a deep thinker who published rarely. In his own words – and university administrators are here warned of offensive material –

> [Cornell] gave me tenure after only two years and with no publication record. In fact, there was one or two papers on the publications list when I was taken for tenure and Francis Low complained that I should have made sure there was none. Just to prove that it was possible.

In terms of realising quantum field theories, condensed matter physics has clear advantages over particle physics. Particle physics is stuck with one universe, and it is not simple to make new ones. There is only one quantum field theory whose behaviour we get to see experimentally – the Standard Model. Now, this is less restrictive than it sounds, because the behaviour of the Standard Model changes depending on the energies of the collisions used to test it and the particles that are collided. Despite being just a single quantum field theory, it can be both strongly coupled and weakly coupled; it can involve both relativistic and non-relativistic physics; it can involve both fundamental particles and emergent composites made from more fundamental constituents.

Nonetheless, the Standard Model is still just one quantum field theory. In contrast, condensed matter physicists can 'make' new quantum field theories by considering different materials, with different impurities, at different temperatures and pressures. Rather than ordering a new universe, they merely need to order materials from the suppliers' catalogue. While this does not allow for an infinity of choices, it does provide an enormous permutation of options.

Condensed matter physics then involves a plenitude of quantum field theories: theories at weak coupling, theories at strong coupling, theories with fermions, theories with bosons, theories with two dimensions, theories with three dimensions, theories with non-relativistic particles and theories with (effectively) relativistic particles. Just as for particle physics, some of these are easier to understand than others. Theories with weakly coupled interactions are preferable to theories with strong interactions. It is both simpler to understand the physics, and quicker to perform the calculations.

Weakly coupled quantum field theories underlie some of the biggest successes of condensed matter physics. For example, most examples of

---

[15] Other fascinating and more recent examples are those of topological quantum field theories, which were developed by Edward Witten in the 1980s. These arise in condensed matter physics to describe states with effective fractional electric charge, as happens in the fractional quantum Hall effect.

superconductivity can be described in this language. Superconductivity is the phenomenon by which a material loses all resistance to the passage of electrical current. In 'normal' superconductivity, a material with few impurities is gradually cooled down. The temperature is reduced towards absolute zero. Once it gets below a certain critical temperature, it undergoes a sudden transition whereby its electrical resistivity vanishes. Current can flow freely without dissipating any heat.

Within a superconductor, electrical current can flow continuously for thousands of years without dissipating energy. Wow – what an amazing world, that has such features in it! Why does this happen?

The cleanest route to the physics of superconductivity is through quantum field theory. A quantum field theory can be written down that describes the behaviour of electrons and atoms in a material with the ability to superconduct. By studying the interactions between electrons and their resulting tendency to pair up, it is possible to see why and when superconductivity occurs. The field theory also allows a quantitative demonstration of the unusual properties of superconductors such as the absence of any internal magnetism and the vanishing of electrical resistance.

Normal superconductors were discovered in 1911 by the wonderfully named Dutch physicist Heike Kamerlingh Onnes. The transition temperature below which superconductivity occurs varies, but is not higher than around thirty degrees above absolute zero. For a long time, this also remained the experimental situation. Following the theoretical explanation of superconductivity in the 1960s, the study of superconductivity appeared to be entering a slow and gentle decline.

However, this all changed utterly in 1986. Out of the blue and inside the reagents cupboard, new superconductors were suddenly discovered with transition temperatures up to three times higher than those previously known – the so-called 'high temperature superconductors'. This electrified the subject, and the strength of the shock is illustrated by the fact that the Nobel Prize for this discovery was awarded only one year later, in 1987.

It is still not known what is the underlying theoretical cause of high temperature conductivity. This problem remains one of the biggest open problems in condensed matter physics. However, whatever the answer is, it is not a simple weakly coupled field theory. Perhaps, the answer requires a quantum field theory that is intrinsically strongly coupled.

There are also other systems in condensed matter physics that involve strong coupling: for example those involving strongly correlated electrons or heavy fermions. In a heavy fermion material, electrons interact so strongly with the magnetic fields of the atoms that they behave as if they were one hundred times heavier than they actually are. The magnetic fields of the material restrict the movement of electrons, increasing the 'effective' mass of the electron. Heavy fermion systems can also lead to superconductivity – but again this is a form of superconductivity that is very different from the 'standard' type.

It is relatively easy to 'make' strongly coupled quantum field theories in condensed matter physics. As we have seen, one of the biggest attractions of the AdS/CFT correspondence is the way it provides a different approach to thinking about strongly coupled quantum field theories, and a different approach to calculating with them. One of the main directions of AdS/CFT research over the last seven years has been to think about applications to condensed matter physics, starting with a 2008 paper by Sean Hartnoll, Christopher Herzog and Gary Horowitz, from a mixture of Santa Barbara and Princeton, explaining how to describe superconductors within the AdS/CFT correspondence. I had been aware of Sean as an undergraduate and graduate student at Cambridge, where he was two years ahead of me, and his untidy long brown hair concealed the smartest brain I was aware of.

Since then, close to a thousand papers have been written on and around this topic, and it represents one of the most active areas of research within this field. There are both reasons both for caution and for optimism.

What are the reasons for caution? Most obviously, the field theories for which calculations can be done using AdS/CFT are not the field theories that actually apply in condensed matter physics. The AdS/CFT theories are normally extensions of the theories that arise in the Standard Model, only with more particles and more symmetry. They are relativistic, and different in many ways from the quantum field theories that arise in condensed matter physics. While something similar also held for the case of the quark-gluon plasma, in this case the differences are starker. For applications to condensed matter physics, it is possible that the only aspect of the AdS/CFT theories in common with the real condensed matter theories is the presence of strong coupling.

At first sight, this appears a total disaster for this research program – calculations can only be done for theories very loosely connected to those generating the actual experimental data. In the end, it may indeed turn out to be a fatal problem.

However, there are also reasons to be optimistic. For any problem, good intuition is extremely important. It is far easier to intuit how a system will behave if you can view it the right way and from the right perspective: once you realise an object has wheels, it is unsurprising when it starts to roll. Most intuition about quantum field theory has been built up through studying examples of weakly coupled theories. While this is fine for studying weakly coupled theories, it may not offer any assistance for strongly coupled examples – and may even be positively misleading.

However, precisely what the AdS/CFT correspondence does offer are calculational methods that apply for large numbers of strongly coupled quantum field theories. These examples allow a development of intuition for the behaviour of such theories, both in terms of the results for the field theory and also by looking at it as a gravitational theory. Even if the calculations are done with the 'wrong' theories, it is a reasonable hope that the wrong theories can still provide insights into the characteristic behaviour of strongly coupled field theories, and that this characteristic behaviour will also carry over to the real theories that describe real materials. If you have always lived on the plains of the Midwest and want to go climbing in the Himalayas, a detailed map and guide of the Alps would still be useful. None of the peaks are exactly the same, but it would show you something of what mountains are like.

The use of AdS/CFT can also offer a guide as to which properties of a quantum field theory are generic, and which only occur within a weakly coupled theory. There are some results which are universal for all quantum field theories, and some which are only a feature of weak coupling. The existence of computationally tractable examples of strongly coupled theories is then enormously helpful, as calculation can be used to show which features survive the transition from strong to weak coupling. Here the benefits of duality enter crucially – it relates theories in which calculations are easy to theories in which calculations are hard. It is only the presence of a dual weakly coupled gravitational theory that allows the relatively rapid performance of calculations in the strongly coupled field theory.

Another danger that does exist comes from an inability to make definite predictions. The disconnect between the theories where one can calculate and the actual real theories describing any given material can lead to a loosening of standards. For this research program, it is both an advantage and disadvantage of condensed matter physics that there are enormous numbers of materials out there. An advantage is that there are lots of targets to study. The downside is that if a calculation in the dual theory agrees with *some* material, it can be taken as a success – but if it disagrees then the mismatch can be explained away as a result of necessary approximations. However there is no systematic way of analysing how accurate the AdS/CFT models are expected to be, and no way of producing an error bar on the theory calculations compared to real experiments on real materials. In this respect, some of the claims made within this area have involved a degree of hyperbole.

The ideal dream within this line of research would be to produce a genuine dual description of a real material. One could then predict the behaviour of a laboratory material using gravitational calculations. While dreams are often unattainable, a more practical motivation is to understand the many different types of behaviour that strongly coupled field theory can lead to. String theory is attractive here as it provides new approaches to think about strongly coupled systems.

In the best of all worlds, this may result in the equations of general relativity predicting new states of exotic matter that can be observed in the lab. On the other hand, this research may end up only reproducing aspects of condensed matter physics already known for a long time. Which is true? How will this research be viewed in thirty years? We simply do not know. That is why it is called research.

CHAPTER 9

Why Strings?
Mathematics

Physicists secretly know their subject to be superior to any other empirical science. It is engrained into the culture of the subject that physics and physicists can bring insights into other sciences, but that the reverse is not true. Physicists become biologists, but biologists do not become physicists. There are master's degrees available for moving from undergraduate physics into the life sciences, but not for moves in the opposite direction. It is also understood *in pectore*, even if never to be uttered explicitly, that the reason for this is that physicists are just, well, smarter.[1] Why, after all, did the discovery of DNA, the most important discovery of twentieth century biology, take place within a physics laboratory? As physicists, we are the best. We have the right stuff. We are the elect of every nation. We are Sheldon from *The Big Bang Theory*.

9.1 PEER RESPECT

There is one subject which is exempted from this benign condescension. That subject is mathematics. Mathematics is, together with astronomy, the most ancient of the sciences, with origins that predate recorded history. It is not an empirical science. The truths of its statements are not contingent on the results of any experiment carried out in this world. They are instead accessible to pure thought, starting from well-defined premises and moving to well-defined conclusions. In the language of philosophy, the truths of mathematics are *analytic* rather than *synthetic*.

For illustration, each of the following statements can be shown to be correct, without any need either to construct measuring apparatus or to perform observations.

---

[1] As the waggish statement goes, physicists used to be both smarter and more arrogant than biologists – now they are just smarter.

1. There is no largest prime number.

2. Any map of countries can be filled in using four colours only, such that no adjacent countries have the same colour.

3. There is no way with only a straight-edge and compass to construct a square with the same area as a given circle.[2]

While mathematics does not require the natural world, much of mathematics is still inspired by the natural world. What makes for an interesting mathematical result is a matter of taste, but taste often prefers a starting point with at least a loose connection to the world of experience. The most successful mathematics book ever written, and also the most successful textbook ever written, is Euclid's *Elements*. Written by the Greek geometer Euclid in Alexandria around 300 BC, this textbook about classical geometry was in continuous use for over two millennia as a workhorse, torment and inspiration for centuries of schoolboys.

The *Elements* include as a key postulate the statement that parallel lines do not intersect. This is true for spaces such as the surface of a table, and these flat spaces are sometimes now called Euclidean. It is not true in spaces that are not flat – for example, the surface of the earth considered as a whole, or the curved spaces that appear in general relativity. In these spaces, parallel lines do intersect, and Euclid's fifth postulate is false – and so then are all the many conclusions in the *Elements* that rely on this postulate. Euclid's postulates were chosen as starting points because in many ways they appeal to our natural intuition. This is one reason Euclid's mathematics has had such lasting value; it is the mathematics suited to spatial geometry as we experience it.

The history of mathematics and the history of physics are not stories that can be recounted independently. There has been cross-pollination between the two to the mutual strengthening of both subjects. The problems of one subject can sometimes be fruitfully expressed in the language of the other, acting as a stimulus to development.

Early examples of problems that sit in both mathematics and physics are the problems of infinitesimals. These are often called Zeno's paradoxes and are the philosophical paradoxes of classical antiquity. One of the best known of these is the question of how Achilles can catch a tortoise. The famously swift Achilles is racing after a tortoise. But how can Achilles ever catch it? For when Achilles has reached where the tortoise once was, the tortoise has

---

[2]This does not stop people from trying to disprove these statements. I was a graduate student in the maths department in Cambridge, which maintained a pigeonhole for the many such crank proofs that got sent in. Anyone was welcome to read, and indeed reply, to the letters in the pigeonhole. One was particularly touching. The envelope was rich to the touch, and inside there was fine craft paper, beautifully and neatly covered with precise writing in an East Asian script. I could read not a word of it – but I knew what the carefully drawn diagram involving a circle and a square meant, and this showed that the neat writing was beyond redemption.

moved on. And when Achilles has reached the place the tortoise is now, the tortoise is another small amount further ahead.

This question is a paradox rather than a genuine problem because we all know that in the real world Achilles indeed overtakes the tortoise. The difficulties lie in correctly accounting for infinitely many infinitely small steps occurring in infinitely many infinitely small units of time. A precise resolution of the paradox requires a precise formalism for dealing with these infinitesimals, which had to wait until the development of analysis ('analysis' here is the technical name of a branch of mathematics) in the early 19th century by the French mathematician Augustin Cauchy.[3] This provides a rigorous treatment of such infinitesimals and how to sum them, thereby dissolving the confusions of Zeno's paradox.

Another example concerns Newton's development of calculus. This was undeniably an enormous achievement in mathematics: to this day calculus remains a *pons asinorum* across which many high school students slowly struggle. However, this development was done not as an abstraction, but as a product of his grappling with the messy problem of the motion and orbits of planets. Newton did not actually include calculus when writing out these results in the *Principia*. He was writing for a wider audience, and he laboriously expressed his results using the classical tools of Euclid that they would have been familiar with. However underlying all this was the slick and powerful machinery of the calculus, which had made all of Newton's calculations vastly easier. While the problem of planetary motions is clearly a physics problem, it acted as the trigger to the mathematical development of calculus.

Moving to living history, a more recent example of the interplay between mathematics and physics occurred during the development of the quantum field theories that describe the strong and weak forces. It was realised in the 1970s that the equations being used here had natural interpretations in terms of the geometric structures which were being studied in mathematics at this time. The discoveries in geometry made in the 1960s and 1970s, led by people like Michael Atiyah, had natural applications to the physics of the strong and weak forces. Atiyah is a remarkable Lebanese-Scots product of a vanished society, having grown up under the British empire first in Khartoum and then in Alexandria. Beyond his outstanding research achievements, he was and is a mathematical panjandrum, ending his career with a list of honours that would have intimidated even Pooh-Bah, with the Fields Medal (the highest award in mathematics), the Presidency of the Royal Society and the Mastership of Trinity College, Cambridge being merely the most notable. The work of Atiyah and others merged the mathematics of the solutions of differential equations

---

[3]Cauchy was born one month after the French Revolution and over the next seventy years lived through all the subsequent vicissitudes of France. One of the greats, his eight hundred papers covered all of mathematics and mathematical physics. Catholic and royalist, he had little sympathy towards the revolutionaries and their successors, who had little sympathy for him, leading to him being rejected for the mathematics professorship at the Collège de France in favour of the serial kleptomaniac and mathematical non-entity Guglielmo Libri.

with the mathematics of topology – the branch of mathematics that explains what makes the surface of a bagel fundamentally different from a sphere, no matter how much you try to squeeze and press the bagel.

What had this to do with physics? As discussed in chapter 5, the 1970s was the decade in which the correct quantum descriptions of the strong and weak forces were determined. During this decade, work in this area transitioned from trying to *discover* the correct theory of the strong and weak forces to trying to *understand* the correct theory of the strong and weak forces. As experimental support for these theories grew, so proportionately more effort went into obtaining a deeper understanding of how to calculate in them and of what these calculations meant.

Both the strong and weak forces involve generalisations of the electric and magnetic fields. Any quantum mechanical calculation in these theories is performed, roughly, by adding up contributions from every possible configuration of the fields – genuinely *every* possible configuration. No matter how bizarre a field arrangement may appear, it still contributes. Certain configurations are more likely, and are weighted more strongly, but every possibility counts. The rules for how to weight different configurations were worked out by Richard Feynman, and this method is called Feynman's path integral approach to quantum mechanics.

It was realised in the middle of the 1970s that certain rare field configurations caused processes to happen that could not occur in any other way.[4] As one example, the proton – the particle out of which we are all made – becomes unstable in the Standard Model. Given enough time, it will decay to other particles. Proton decay is the ultimate ecological catastrophe – without protons, there are no atoms, and without atoms, there is no life. Fortunately, the values of the constants of the Standard Model happen to be such that not a single proton in our bodies would have decayed even after waiting many lifetimes of the universe.

What made these configurations special, and what distinguished them, was their topology. They involved arrangements of fields that, no matter how much you tried, no matter how much you bent them, you could never smoothly deform to a zero configuration. When did these special configurations contribute? What were the conditions? The answer was given precisely by the work of Atiyah and his collaborator Isidore Singer, in particular through the Fields Medal-winning theorem they shared, the Atiyah-Singer Index Theorem.

There is a second example from the same time period of the interplay between maths and physics. It comes from a famously productive 1960s collaboration between the physicist Stephen Hawking and the mathematician Roger Penrose. Hawking was trained as a physicist and has since become an iconic image of science; Penrose was trained as a mathematician and came from a distinguished polymath family. Penrose's father was a noted psychiatrist and his brother was British chess champion.

---

[4]Technically, these are called *instantons* or *sphalerons*.

Both Hawking and Penrose were interested in geometry, and in particular in the geometries that arose within general relativity. They were interested in questions such as: what happens when a blob of matter starts collapsing under gravity? It had been known since the 1930s that the collapse of a perfectly symmetric shell of matter leads to a black hole. However it was unclear what happened for more general circumstances. Did all the collapsing matter fall into a black hole, or could some of it be expelled via some kind of slingshot mechanism? Similar questions arose in the early history of the universe. If the expanding universe we observe today was extrapolated back in time so that it became progressively smaller and progressively denser, did the equations of general relativity necessarily lead to a singularity where infinities appeared and general relativity broke down?

Together, Hawking and Penrose solved these problems and showed that singularities were unavoidable. They did so using novel techniques that involved looking at the global geometry of spacetime, rather than simply the local behaviour of small regions. By looking at spacetime as a whole, Hawking and Penrose could see what caused what and what followed what. By applying these geometric techniques to general relativity, they were able to map out the entire picture of the geometry – and by doing so they were able to show that singularities were an unavoidable feature of classical general relativity. As the universe was extrapolated back in time, Hawking and Penrose showed that the equations signalled their own demise, and led inexorably to a regime where they ceased to be valid.

These and other applications rekindled the long dormant *affaire de coeur* between physics and mathematics. The physics that had triumphed between the 1930s and the 1960s was not mathematical in nature: the mathematics used did not come from the front line, and the development of physics and the development of mathematics were proceeding on separate paths. Richard Feynman may have been a great physicist, but he and his generation left no direct trace on mathematics. They were physicists first and physicists second, and they successfully explained a succession of new data without requiring mathematical sophistication or results that would impress their colleagues in the next department. The 1970s and 1980s were different: for the first time in a generation, modern physics and modern mathematics were sparking off each other and pushing ideas in each other's direction.

This section has described some examples of the longstanding conversation between mathematics and physics. A good reason for working on physics is because it provides novel ways of looking at mathematical problems and interpreting them. A good reason for working on an area of mathematics is that it involves topics arising in, and required by, modern physics.

These features have been well manifested in string theory. There are many people who work on string theory, or use ideas derived from it, because it can provide either solutions or insights into mathematical problems.

I want to describe two specific examples involving string theory, starting with the wonderful name and wonderful tale of the monstrous moonshine conjecture. This is the story of an interplay between the mathematics of symmetry, geometry, number theory and string theory. It is a story whose first wave lasted for around two decades, and for which smaller wavelets continue to break even now.

9.2   OF MONSTERS AND MOONSHINE

Once upon a time there was a beautiful symmetry that almost no one cared about. As we have seen in chapter 3, the basic notion of a symmetry is an operation that, performed on something, brings it back to itself. We saw there examples of rotational symmetry such as fourfold symmetry or sixfold symmetry. It is easy to see how this can generalise to arbitrarily large examples: hundredfold or thousandfold rotations. There are also other examples of families of symmetries: for example, permutation symmetry, under which groups of objects remain the same when shuffled.

What types of symmetries can exist? Mathematicians like to classify symmetries, and one of the most epic achievements in mathematics has been the classification of 'all possible symmetries'. This is a theorem, technically known as the classification of finite groups, which states what all the possible allowed forms of symmetry are. This theorem was built up over decades through the work of hundreds of mathematicians, covering tens of thousands of pages across different mathematics journals.

This theorem describes every possible form of symmetry that can arise. Some of these forms are well known, such as the rotational symmetries described here, while others are less familiar. However, almost all symmetries can be grouped into one of several large families of related symmetries. We are interested here in the small number of exceptions. These exceptional symmetries have remarkable properties. They are called sporadic symmetries, and they number twenty-six in all.

The sporadic symmetries are not easy to grasp. One of their striking properties is that they are all 'large'. What is meant by 'large'? The symmetry of fourfold rotation is said to have order four, and the symmetry of sixfold rotations is said to have order six. The order describes how many independent ways there are of acting with the symmetry. On this count, even the smallest of the sporadic symmetries has an order of just under eight thousand. This already sounds quite impressive, until you realise that the largest of these symmetries has an order of approximately ten raised to the fifty-fourth power – a billion multiplied by itself six times. This symmetry is, for this gargantuan number of reasons, called the monster symmetry.

How can one possibly understand a symmetry of this size? What can such a symmetry act on? What objects play the role for it that the square does for fourfold rotations? The same theorem tells us that the smallest interesting object that the monster symmetry acts on has 196 883 elements (the uninteresting object is a single element, taken to itself by every action of the symmetry). The action of the monster symmetry turns all these different elements into each another – but it is again hard to get a conceptual handle on anything with almost two hundred thousand components.

The monster symmetry is a symmetry like no other. Its existence was first realised in the middle of the 1970s by the mathematicians Bernd Fischer and Robert Griess, and a few years later Griess was able to give an elaborate and explicit definition of the monster symmetry in terms of its actions on this object with 196 883 elements.

As an unusual and large special case amidst the general taxonomy of symmetries, the monster symmetry might easily have remained only as an odd member of a mathematical curiosity cabinet. It did not, and the reason it did not starts with a numerological observation by the Canadian mathematician John McKay. McKay had originally worked on number theory, the branch of mathematics that deals with such questions as the nature and distribution of prime numbers. One function that appears within number theory is called Klein's j-function. This function has been known for a long time and is named after Felix Klein, a nineteenth century German mathematician. It obeys many striking relationships, and for our purposes there is one particularly useful way of writing this function. This is as an expansion with a series of coefficients, the first of which are 1, 196 884 and 21 493 760.

What was McKay's flash of genius? He noted that

$$196\,884 = 196\,883 + 1.$$

Of course, this equation is trivially true. McKay's hunch was that the closeness of these two numbers was indicative of a unifying structure connecting the monster symmetry to properties of the j-function. At first, this looked like it was as meaningful as a relationship between the value of π, the fine structure constant, and the volume of the Great Pyramid of Giza when measured in cubits. If you look at enough numbers from enough places, there will always be some odd coincidences that appear. The possibility that this actually represented a real connection – that seemed pure moonshine.

It soon became clear that it was not. McKay told another mathematician John Thompson about his observation. Thompson was a mathematical giant who had already won the Fields Medal for previous work on the classification of symmetries. Thompson took it seriously, and discovered another coincidence. The next coefficient of the j-function was 21 493 760, while the second-smallest object that the monster symmetry can act on has a size 21 296 876. Thompson observed that

$$21\,493\,760 = 21\,296\,876 + 196\,883 + 1.$$

It was not just the first coefficient of the j-function that had connections to the monster symmetry, but also the second one – and Thompson checked that there were similar relations that held for the next few coefficients as well. It was now clear that McKay's observation was no fluke, and that there did exist a deep underlying relationship between these two disparate areas. These and other 'coincidences' were summarised in a 1979 paper by John Conway and Simon Norton entitled 'Monstrous Moonshine'. The challenge was clear: explain why these relationships existed and where they came from.

The person who ultimately solved this problem was Richard Borcherds. Borcherds is a British mathematician who has now moved to the United States. As is not uncommon for mathematicians, he was also a strong chess player as a child, before moving on to more productive intellectual pursuits. His work on this problem would see him awarded the Fields Medal. While the language of Borcherds's proof was the language of mathematics, the ideas are interleaved with string theory and in part draw directly on key results from it. What follows is an unabashedly physics summary of this work.

The string theory relevant for Borcherds' work was the simplest string theory: the bosonic string. This was the first string theory to be constructed and has the smallest number of complications. The bosonic string theory is not viable as a description of the world – as was mentioned in chapter 5, it contains a tachyon particle that renders the theory unstable and unphysical. However, for mathematical applications this instability is of no concern.

The bosonic string needs twenty-five spatial dimensions for consistency. If you twang it, there are twenty-four separate directions in which to pluck it – all dimensions except the one along the string. There are many ways to pluck the string, and the string has many harmonics. These harmonics can be counted. In the simplest case where all twenty-five spatial dimensions are identical, flat and infinite in extent, the first harmonic has a multiplicity of twenty-four, associated to all possible one-pluck notes. The second harmonic can be obtained either through simultaneously plucking the string along two directions, or by giving it an extra-hard 'double pluck' in one direction. These two possibilities can be done in many more ways, and so the multiplicity of the second harmonic is greatly increased. This is a verbal description, but mathematical formulae can easily be given to count the multiplicity of harmonics up to arbitrarily high levels.

The functions that count the harmonics also make appearances in number theory. The precise function that appears in any one case, however, depends on the geometry of the spatial dimensions. Unsurprisingly, the ways that strings can be plucked is affected by the geometry of the space that they live in. For the more conventional string theories these functions have names such as the *Dedekind eta function* or the *Jacobi theta function* – classical functions from nineteenth-century number theory.

Where does monster symmetry come in? We are already familiar with the fact that spatial geometries can have symmetries – for example, the square has fourfold symmetry. If the geometry of string theory's extra dimensions involved a square, that fourfold geometric symmetry would also be a symmetry of the ways that strings could be plucked. Given a single way of plucking a string, a fourfold rotational symmetry would automatically generate three more. The symmetry of the geometry would be inherited by the symmetry of the strings, and a count of the multiplicity of harmonics would have to respect that symmetry.

The key idea of Borcherds was to identify a geometric space that had the monster symmetry as its symmetry, and then use the geometry of this space for string theory's extra-dimensional geometry. The space Borcherds used was a variation on a particular twenty-four-dimensional space called the Leech lattice, which had previously been recognised as having connections to the monster symmetry. Instead of doing bosonic string theory on the simplest spaces, Borcherds was doing bosonic string theory on this bizarre Leech geometry – but a bizarre geometry that was governed by the monster symmetry.

The fact that the monster symmetry applied to the geometry meant that the monster symmetry also applied to strings forced to vibrate within it. A string governed by a symmetry has to dance according to its tune, and when the symmetry is the monster symmetry the tune is that of the monster. Any way a string vibrated, the symmetries of the monster could turn the string into one vibrating another way. The rules of the monster constrained the allowed vibrations of the string – they had to come in groups of 1, or 196 883, or 21 296 876, or larger. The mathematics of monster symmetry allowed no other options. It was here that the real power of the symmetry entered, by enforcing these highly distinctive numerical groupings of the allowed vibrations of the string.

Borcherds was also able to determine the function that counted the string harmonics for this geometry. In the simplest versions of the bosonic string, this function had been the Dedekind eta function. For Borcherds' carefully chosen geometry, it was now the Klein j-function. Written as an expansion, each progressively higher coefficient of the j-function was now counting the multiplicity of the string harmonics. The coefficient 196 884 was the number of realisations of the 'first harmonic', the coefficient 21 493 760 the number of realisations of the 'second harmonic' – and so on. This gave a 'physics' meaning to the j-function: it was counting harmonics of the bosonic string on the Leech lattice. The fact that these coefficients could all be expressed as neat sums of the sizes of monstrous objects reflected the monster symmetry that acted on the geometry. The harmonics *had* to arrange themselves in this way, since the oscillating strings were forced to come in symmetry groupings. As the applicable symmetry was the monster symmetry, the sizes of these groupings had to be those of the monster.

Through this route – although expressed in a far more mathematical language! – Richard Borcherds was able to prove the statements made in the monstrous moonshine conjectures, and for this work he was awarded the Fields Medal at the 1998 International Congress of Mathematicians in Berlin. For each award, there is a Laudation – a formal speech praising the winner and describing their work. As part of his proof, Borcherds had made crucial use of the No-Ghost theorem from string theory, which was mentioned in chapter 5 and had been proved by Peter Goddard and Charles Thorn in 1973. As a result, it was Peter Goddard who was invited to given the Laudation. The invitation left Goddard feeling 'enormously flattered, but faced [with] the most formidable challenge'. The language of formal mathematics is often difficult even for the best physicists.

This story of monstrous moonshine is then an example in which string theory provided some of the stones for a bridge connecting two disparate areas of mathematics. It is not yet a finished story. In recent years further examples of moonshine have been found, which connect some of the other sporadic symmetries that are tamer versions of the monster symmetry to different number-theoretic functions. These examples are not yet understood, but it is expected that it is string theory on a particular special geometry that will explain the connections. Teams of physicists and mathematicians are even now working together to uncover the string theories that can explain these links.

9.3   MIRROR SYMMETRY

The first mathematics we all encounter is that of counting:

> One potato, two potato, three potato, four;
> Five potato, six potato, seven potato, more!

I have an uncle who is a professional mathematician. When I was a very small child, in that I had any sense of his work, I thought it involved counting really, really large numbers. As I grew slightly older, I matured in my views. No longer did I think Uncle Joe's work simply involved counting. I was now at school, and learning to multiply two-digit numbers together. What he did, I now imagined, would be the grown-up version: multiplying twenty digit numbers together, and similarly performing by hand labyrinthine long division calculations to many decimal places. If primary school mathematics was about sums, grown-up mathematics would be about hard sums.

As a child I was both wrong and right. Of course, in many ways I was wrong and immature. Real mathematics involves deep structures and arguments held together by careful chains of reasoning. While some parts of it, such as the proof that there is no largest prime number,[5] are accessible to children, most is, like the drinks cabinet, out of the sight of young eyes. However, I was not entirely wrong. Some parts of grown-up mathematics – although not in fact the area my uncle worked on – do involve counting. They involve working out the total number of ways that something can happen. For example, how many ways are there of writing a whole number as a sum of smaller numbers? Two can be written as either two or one plus one. Three can be written as three, two plus one, or one plus one plus one. How does the possible number of sums

---

[5]To see this, suppose that there was actually a largest prime number and thus a finite set of prime numbers. One can consider the number made by multiplying all of the prime numbers together and then adding one. A little thought convinces you that this is also a prime number, contradicting the original assumptions – and therefore the supposition that there is a largest prime number is mistaken.

grow as our original number gets very large? This is a counting problem, and it is the simplest form of many such related problems in mathematics.

The theory of numbers is one area where counting problems appear in mathematics. Another area is in geometry. While mathematicians' usage of the word 'geometry' can be broad, I will confine its meaning here to that of regular spaces. Although they may have two, three, four, five or six dimensions, these spaces are just fancy generalisations of familiar circles, spheres and ellipsoids.

What do geometers count? One answer is curves – how many different curves are there in a geometry? Equivalently, how many ways are there to wrap an elastic band?

What does this mean? As a more precise way of formulating this problem, we imagine an elastic band – or a string, or indeed any object with tension – that is confined to a surface and is not allowed to move off the surface. The surface could be the surface of a sphere, a cylinder or a bagel (more technically known as a torus). We imagine placing the elastic band on the surface, and asking what can happen to it. The number of possible results depends on the surface – and this is what mathematicians count. For a sphere, the band always shrinks down under tension to zero size. It has nothing for it to wrap itself around and so can never prevent itself from collapsing. For a cylinder, the band can however wrap around the cylinder – either once, or two, three or four (or more) times. Provided the band has to remain on the surface and cannot be slipped off, there is no way to untangle these wrappings. For the bagel (or torus), there are now actually two distinct ways of wrapping the band, one for each of the circles present in the bagel. Each of these circles may again be wrapped either once or many times.

This counting problem has different answers for spheres and tori. The answer also remains the same however one kneads or moulds the surface. If a sphere is deformed into an ellipsoid, the elastic band can still shrink itself to zero size. However a band wrapped around a cylinder remains wrapped – and cannot be unwrapped, even if we squeeze the cylinder. This is useful information, as it shows that this counting problem provides a way of distinguishing surfaces from one another. As the answer does not depend on how one deforms the surface, it depends only on the *topology* of the surface – and therefore surfaces for which the count is different are topologically distinct from each other.

I have introduced this geometric counting problem with one-dimensional elastic bands. More generally, this mathematics is the mathematics of counting the number of possible ways that a circle, or the surface of a sphere, or the surface of a hyper-sphere, can fit into geometries with additional dimensions. For any given geometry, one can ask how many 'different' ways there are to put one of these objects in the geometry. This branch of mathematics is called *homology*, and in a loose sense it counts holes in surfaces.

It is interesting because it provides a way of distinguishing different geometries and quantifying their difference. As the geometries grow more complicated, so do the techniques. While finding the possible ways of putting an elastic band around a circle may not be taxing, counting the number of ways of embedding a sphere in a six-dimensional Calabi-Yau geometry requires considerable thought and training.

Calabi-Yau spaces were first mentioned in chapter 5. They are a type of geometric space that can have any even number of dimensions. They arose in the 1970s as a type of geometry of interest to mathematicians, and in particular to those who were algebraic geometers. As mentioned in chapter 5, it was realised in 1985 that six-dimensional Calabi-Yau spaces are promising starting points for particle physics models descending from string theory. Provided the additional dimensions of heterotic string theory are curled up into a Calabi-Yau space, the resulting four-dimensional theory has rough similarities to possible extensions of the Standard Model. Indeed, as seen in chapter 5 it was this result that first led to applications of string theory to particle physics.

Before this paper, not one physicist anywhere in the world cared one whit about Calabi-Yau geometries and their properties. After it, many did. One of its key results was that the particle content in four dimensions depended on the geometric properties of the Calabi-Yau space – and in particular on the number of possible ways that spheres or hyper-spheres could fit into it. For the simplest class of models, the number of particle generations was determined by the difference between the number of ways of embedding two-dimensional surfaces and the number of ways of embedding three-dimensional surfaces. While in the end this feature does not seem to give useful insight concerning the presence of three generations in the Standard Model, it does explain why Calabi-Yau geometries came under intense investigation by a community trained as physicists – who suddenly cared about counting the number of ways to embed surfaces into such geometries.

Mathematicians tend to be abstract. Physicists like examples. One of the first activities of these physicists was to attempt to construct as many Calabi-Yau geometries as possible. In collaboration with mathematicians, new techniques were sought to increase the number of known Calabi-Yau spaces. With these techniques, the number of such known spaces grew first from a few to tens, then from tens to hundreds and then from hundreds to thousands. As the numbers grew, it became possible to examine the properties of this large collection of geometries.

As this was done, an interesting property was encountered. In technical notation, the numbers used to count the ways of embedding two-dimensional and three-dimensional surfaces are called $h^{1,1}$ and $h^{2,1}$. These numbers can reach from unity up to values of almost a thousand. However, a curious pairing was noticed. For example, there was a space for which $h^{1,1}$ was one and $h^{2,1}$ was one hundred and one. However there was another space for which $h^{1,1}$ was one hundred and one and $h^{2,1}$ was one. As more and more Calabi-Yaus were enumerated, it was noticed that this pairing structure was ubiquitous. If all known values of $h^{1,1}$ and $h^{2,1}$ were plotted on a plane, the result looked as if it had been reflected about a central mirror. For every Calabi-Yau space, there appeared to be another mirror partner space for which the values of $h^{1,1}$ and $h^{2,1}$ were interchanged.

These mirror pairs exhibited a further striking feature. It was realised that the equations obtained from starting with type IIA string theory, and compactifying it on one Calabi-Yau space, were exactly identical to the equations obtained from starting with type IIB string theory and compactifying it on the mirror partner of the original space. This was another example of a duality – one theory on one geometry was identical to a second theory on a different geometry.

It is this last fact that led to the reason why mirror symmetry became of such interest to mathematicians. As with other dualities, it is also true here that quantities that are hard to compute on one side of the duality become easy to compute on the other. Quantities that were hard to compute for IIA string theory on one Calabi-Yau were 'easy' to compute for IIB string theory on the mirror space. In particular, it was realised that the 'easy' classical computations on one side were equivalent to full quantum calculations on the other side, including both perturbative and non-perturbative effects.

These full quantum computations included quantities of clear mathematical interest. In particular, they contained a count of what can be roughly described as the number of ways of writing curves within the Calabi-Yau space. This is a similar but distinct quantity to the number of ways, alluded to above, of fitting surfaces and hyper-surfaces within such a space. For the simplest such curves, said to be of degree one, this count had been performed in the nineteenth century. For more complicated curves of degree two and degree three, the count had only been performed more recently – but the techniques were complicated.

What was exciting about mirror symmetry was that the string theory calculations provided an expression for the number of curves – not just of degree one and two, but of degree four, five, six and higher. In one go, it produced an expression that purported to count the number of curves in a Calabi-Yau space up to arbitrarily high degree. Philip Candelas, then of Texas and now of Oxford, presented these results to a conference of mathematicians in Berkeley in 1991. For the simplest example the numbers went as follows:

Degree	Number of curves
1	2 875
2	609 250
3	317 206 375
4	242 467 530 000
5	229 305 888 887 625

While the results agreed with the existing mathematical results for curves of degree one and two, they disagreed for curves of degree three. For a few weeks there was an impasse – but then the mathematicians who had obtained this result rechecked their codes and found a bug. The new result agreed precisely with the physical result presented by Candelas.

This made it clear that string theory was both able to say deep things about Calabi-Yau spaces and also able to obtain results about them that were inaccessible to conventional mathematical techniques. This was exciting both to mathematicians and physicists – and by now has led to an enormous amount of work in this area, going far beyond these original results. Mathematicians started learning string theory and trying to understand its tools and methods – because it was able to tell them new results about geometric spaces they were interested in.

9.4   CULTS IN PHYSICS

When considering the influence of string theory and related ideas on mathematics, there is one name that comes up more than any other. That name is Edward Witten of the Institute for Advanced Study in Princeton. Witten has become an icon of the subject, despite only settling on a career in physics after a meandering start, including an undergraduate major in history, a foray into journalism and dropping out of an economics doctoral program after one term. He went to Princeton and started as a graduate student of David Gross. He was in fact the second graduate student of Gross, the first being Frank Wilczek, whose Nobel Prize-winning doctoral work featured in the introductory chapter.[6] Witten was thus one of the first members of that generation whose entire careers have been subsequent to the construction of the Standard Model.

Witten has made numerous contributions to quantum field theory, general relativity, string theory and mathematics. Witten's contributions to modern mathematics have been sufficiently numerous that he was awarded the Fields Medal in 1990 – with the Laudation given by Michael Atiyah – thereby becoming the only physicist so far to win mathematics' highest award. Various results led to this prize. He had shown how the theory of knots – what are the different knots you can make with a piece of string? – is related to quantum field theory, and how some of the interesting properties of knots can be computed using quantum field theory. He had provided a remarkably simple proof of the fact that any gravitational system has positive energy. He had pioneered an entirely novel approach to Morse theory, a branch of topology – and all these were only the achievements in his second subject.

When I was a student, there was almost a cult about Witten. The name was mentioned with slightly hushed tones. He was someone on a different plane, perhaps even from a different planet. Students who had attended summer schools at Princeton would return and report on exactly what he had said when asked questions in his talk, and the precise words he used in answering. It was joked, only partly in jest, that he was a member of a superior alien species masquerading as a human. The unusual delivery for such a tall man,

---

[6]With two such predecessors, one can only feel sympathy for the third graduate student of Gross!

with a high-pitched, slightly squeaky voice – an oracle speaking as a counter-tenor – only reinforced the sense of difference. He had the ability to make, or break, an area. What he worked on was automatically fashionable – if he thought something was important, it was. There were many other smart people in the subject, but there was only one Big Ed.

The collective memory contained stories of scientists who had made their careers during the 1980s by chasing after anything Witten had written. In physics, research articles are circulated in preliminary form prior to publication, originally through paper copies and later via electronic mail. Once such a 'preprint' appeared with Witten's name on it, these scientists would read it, digest it, and then quickly write a follow-up paper on the same topic – take what Witten had done, make a small perturbation, and re-solve for a different but closely related example. The final step was to send the hurried composition off to a journal; and then await the next preprint to repeat the same trick.

How did this arise? Why was one man held in such awe? Part of the answer is certainly Witten's prodigious talents and ability. However these talents also had a rich field to harvest. As mentioned previously, the particle physics from 1945 to the middle of the 1970s was not opposed to mathematics, but it was also not really interested in mathematics. The development of the Standard Model had stimulated great advances in the tools and theories available to particle physics. The great names of the age cared more, though, about pursuing with vigour experimental data and its possible explanations. A full understanding of the deep structure of the equations they came up with was not what truly excited them.

What this meant is that by the late 1970s the area at the interface of mathematics and physics lay entirely ready to be exploited by someone with the right set of talents. Edward Witten was exactly that person. He had the ideal combination of abilities. Few physicists have a similar command of mathematics, and few mathematicians can match his deep physical insight. Witten has been extraordinarily productive at the area at the boundary between mathematics and physics. He was someone who was enormously gifted – and also someone whose talents peaked in an area that had lain fallow for decades and was now fertile soil for those with the ability to cultivate it. In all areas of science, the first to the orchard gathers abundant fruit from the ground and from the low-lying branches, while those who arrive later need elaborate ladders for much smaller pickings.

Witten's achievements and influence have been such that for a long time he represented the model of how to do string theory, and even to a certain extent the model of how to do theoretical particle physics. Sheldon Glashow, one of the architects of the Standard Model, a 1979 Nobel Prize winner, and a trenchant opponent of string theory throughout the 1980s and 1990s, wrote a poem that ended,

> Please heed our advice that you too are not smitten –
> The book is not finished, the last word is not Witten.

Glashow had attempted to keep string theorists out of Harvard, where he was a professor. He did not succeed; he left Harvard and has now moved to nearby Boston University.

Cults have their dangers. The dangers of a scientific cult is that the model of the 'ideal scientist' can remain the same even while the science shifts, and the skills required for the next big breakthrough change. A previous cult had existed around Richard Feynman – the fun-loving, safe-cracking, girl-seducing Dick Feynman described in *'Surely You're Joking, Mr Feynman?'*. The Feynman cult still exists in attenuated form today, and I see it every year in the personal statements dutifully compiled by British teenagers seeking entrance to university. To many, Feynman is still the model – if not of how to behave, at least of how to do physics. Feynman was indeed a great physical scientist – but he was also one whose highly intuitive approach and style rendered it impossible for him to make any of the more mathematically oriented breakthroughs of the 1970s and 1980s. It is not just that Feynman did not happen to be the physicist who got these results. He could not. His scientific style made him constitutionally incapable of it.

Feynman may or may not have been a genius, but if everyone tried to be like him, none of these results would have been found. The same is true of Witten – not every interesting problem belongs at the intersection of mathematics and physics. Different skills are needed for different questions, and no one ever became a principal dancer at the Royal Ballet through years of practising the oboe.

What is the summary of this chapter? The summary is that there are many scientists – some would argue too many – for whom time spent on string theory is time spent for the greater glory of mathematics. These scholars fall into several groups. There are those whose training is in physics, and who in a long-distant past did indeed torture the apparatus in laboratory courses. They use physical reasoning and argumentation to provide, for what are essentially problems in mathematics, novel lines of attack and fresh calculational methods. The benefits of looking at a problem from a new angle are clear, as it reshuffles what is 'obvious' and what is 'hard'.

There are also those who are trained and work as mathematicians. They may take the intuitive style of physical argumentation and toughen it, by alloying to it the rigour of professional mathematics. They seek to understand, in their own way and according to their own discipline, the underlying mathematical structures that enables physical reasoning to find the right answer to mathematics problems.

There is another, slightly looser, class of mathematician. These are those who might use string theory or its structures as a kind of muse, but who are not directly engaged with it. For all scientists there exists an intellectual penumbra of ideas and topics that one knows loosely, and watches, but does not work on actively. Such mathematicians have no professional interest in quantum gravity, and even less in the problems of the Standard Model, which they would be hard pushed to define. However they do care about the mathematics of

Kac-Moody algebras, and techniques to give information about the geometry of Calabi-Yau spaces. For such people stringy mathematics is, even if they do not work on it directly, a part of their wider scientific culture.

The style of work discussed in this chapter has been called 'physical mathematics' by Greg Moore, a physicist-cum-mathematician at Rutgers University in New Jersey. The problems to be solved are ultimately mathematical in nature – the answers are not contingent on any observation or experiment. However the tools used to attack these problems were all forged within theories devised for the purpose of studying nature.

It should be noted that in 'physical mathematics', 'physical' qualifies 'mathematics' – and not the other way round. Such researchers will be happy if the results they discover are eventually relevant to nature. The satisfaction they feel, though, is complete in itself. Their pleasure in their research is not tainted by any unfulfilled yearning for messy data or experimental discovery. Understanding our world is not the province of mathematics, and it is not the metric by which mathematicians rank each other. Mathematics is done for the sake of mathematics and not for utilitarian purposes. The mathematician's mathematician G. H. Hardy famously wrote in a *A Mathematician's Apology*,[7]

> I have never done anything 'useful'. No discovery of mine has made,
> or is likely to make, directly or indirectly, for good or ill, the least
> difference to the amenity of the world.[8]

For Hardy and many others, the value of their subject is not to be found in its ability to address anyone else's concerns.

Scientists with a career work on a combination of what others find interesting, what they find interesting and what they can make progress on. Happiness occurs when all three coincide. For certain areas of mathematics and certain flavours of mathematician, string theory has been the cause of this coincidence.

---

[7]Hardy was for eleven years a fellow of the same Oxford college – New College – of which I am a member, living only a few metres from where I type this. He was a more than usually reserved member of that generation of British men for whom the use of first names represented almost carnal knowledge. His contemporary, the economist Lionel Robbins, described him then as 'No depiction known to me of a saint receiving the stigmata shows greater intensity than did Hardy's features when plunged into meditation. Nor is it easier to conceive a more vivid exhibition of the meaning of the word 'illumination' than was afforded by the same features lit up by the play of wit or intent in kindly badinage. And nothing in the man belied the appearance.'

[8]Despite all his good intentions, Hardy's work on number theory now has numerous applications within cryptography.

CHAPTER 10

Why Strings? Cosmology and Particle Physics

We have seen in the last two chapters various reasons to care about string theory. However, none of these reasons have involved the idea that string theory makes statements that are really true about this world at the smallest possible distances. What makes string theory interesting to those who care about the deepest laws of this world?

This is the purpose of this chapter. I aim to explain how string theory can connect to known physics while suggesting novel ideas for going beyond it. There are several ways to approach this, but as this is what I work on, in this chapter I shall take the indulgence of writing about my own research.

10.1 EXTRA DIMENSIONS AND MODULI

String theory is famous as a theory of gravity, and in particular as a theory of quantum gravity. However we have also seen that to string theory, it is gravity in ten dimensions – and not four – that is the favoured first-born child. In addition to the time dimension, the familiar three spatial dimensions are extended by a further six dimensions.[1] These dimensional appendages are intrinsic features of the theory – the form of gravity string theory gives us is, of necessity, one involving extra spatial dimensions.

There is an advanced side comment here. In certain limits, these extra dimensions cease to behave as a classical space within which familiar interpretations of dimension and geometry can be sustained. In these limits, the behaviour of the extra dimensions is best described as quantum geometry.

---

[1] As we have seen in chapter 5, in some limits of the theory there are an extra seven spatial dimensions, but this embarrassment of poverties will not modify this discussion.

This idea will reappear in the next chapter, but for now I will simply say that this behaviour has a well-defined mathematical description and can be smoothly joined up with the more familiar language of classical geometry. In these limits, string theory can be said to require extra quantum dimensions rather than extra classical dimensions.

Be that as it may, the necessary existence of extra dimensions is at both first and second sight a serious problem. For an idea to be right about this world, it first of all has to be not wrong, and 'wrong' is exactly what an additional six dimensions of space appears to be. Before discussing minutiae, we first need to explain why the existence of extra dimensions, in the real universe, is not in flagrant contradiction with observations. Having done this, we can then explore the idea further, and examine what the positive consequences of extra dimensions are for the physics of this world.

So why is the existence of extra dimensions not immediately marked as inconsistent? We know about two ways of moving across and one way of moving up. There is no 'hyper-up' direction: where is this extra dimension?

To address this, it is helpful to think about what we mean by a dimension and how we distinguish between one, two and three spatial dimensions. The easiest way to think about this is in terms of labels. In a space with two dimensions, we need two labels – two numbers – to uniquely specify the position of an object. Whether we call these labels x and y, or latitude and longitude, or distance and angle, the key feature of a space with two dimensions is that it involves two labels. The analogous property is also true for three or one dimensions: in a three-dimensional space, we need three coordinates to label where something is and in one dimension you just need a single number. We also label positions with distances, and the better our rulers are, the more we can refine our notion of a position. It does not make sense to talk about objects being a micrometre apart if the only ruler we have has gradations of millimetres.

If extra spatial dimensions are present, there must be additional coordinates we can use to describe position. These extra spatial dimensions are consistent provided that all distance scales involved are smaller than the distances our rulers can resolve: the extra dimensions are curled up. There is an analogy for this as good as it is standard: an ant walking along a piece of string (the 'string' here is the regular type one buys from the supermarket!). The walking ant experiences the string as a one-dimensional object. The ant can go forward and it can go backwards. There is only one direction it can move along, and the world of the ant is effectively one-dimensional.

The string does however have another direction: the transverse direction around itself. For a bacterium deposited by the ant's feet, the surface of the string is two-dimensional. The bacterium is so small that even the direction around the string still counts as large, and because it is so much smaller than the ant, its world is two-dimensional.

In a similar way, extra dimensions that are small enough are literally invisible to us. The wavelength of visible light is around one ten-thousandth of a centimetre: any distances smaller than this cannot be resolved by means of visible light. Provided these extra dimensions are smaller than the resolution of any probe available to us, their possible existence is consistent with all observations and experiments performed to date. We can no more tell that they are present than use our fingers to feel viruses. In practice, this requires the sizes of extra dimensions to be less than around one billionth of a nanometre.

What then are the implications of extra dimensions? For if string theory is true, the existence of extra-dimensional gravity is a genuinely true feature of the world – at the smallest possible scales, there are more than three spatial dimensions. If this statement is true, it should have consequences. What are these consequences? As ten-dimensional gravity is the Texan version of Einstein's gravity – the same theory, except bigger and better – the behaviour of ten-dimensional gravity both includes and generalises the behaviour of four-dimensional gravity.

What does this imply? There are many paths one can follow here and I am going to choose one, guided by my own research and interests. I shall explore one aspect of four-dimensional gravity and explain what its generalisation to ten dimensions implies. This path starts with the fact that Einstein's theory tells us that spacetime is dynamical. It is not the rigid globe of the Ptolemaic heavens, unchanged and imperturbable. It is responsive, as the air is responsive to the click of fingers through the production of sound waves. Spacetime itself, and its geometry, sends out ripples in response to a disturbance, just as the surface of a pond ripples with water waves in response to the entry of a stone or a hippopotamus. These ripples in spacetime are called gravitational waves, and their size and strength depend on the magnitude of the disturbance. In the same way that a belly-flopping hippo makes larger water waves than a dropped pebble, mergers of black holes make larger gravitational waves than colliding ping-pong balls. These waves all represent changes – whether macro or micro – in the geometry of spacetime, propagating outwards from their point of origin.

Such gravitational waves have never been directly detected. This is because they are intrinsically very weak and so their observation requires exquisitely precise technology. When direct detection does first occur, it will come from extreme astrophysical events, such as collisions between black holes or massive stars, capable of generating the largest ripples in spacetime. As with any other kind of wave, these ripples become weaker as they spread out. By the time they reach us, the geometric perturbations the ripples represent manifest themselves as distance wobbles as small as one nuclear width across a length of several kilometres. It sounds fantastical that one could attain the technological precision required to measure such a change, but this is precisely the goal of the current and planned gravitational wave experiments that hope to detect gravitational waves within the next decade.

Gravitational waves are a clear prediction of Einstein's theory. Despite the lack of a direct detection, there is extremely strong indirect evidence for their existence, beyond simply the accumulated evidence for the correctness of general relativity.

As with any other form of wave, gravitational waves carry energy. A system currently emitting gravitational waves has less energy after the emission than before, and this change in energy can be measured. This precise effect can be seen in stellar binary systems of a pulsar in a close mutual gravitational orbit with another star. A pulsar is a magnetised, rotating neutron star that emits a beam of radiation at a precise frequency. Every time the star rotates, the beam passes across earth like the beam of a lighthouse. As the pulsar orbits its companion star, gravitational waves are emitted which carry off energy. The energy that is carried off reduces the orbital energy of the system, bringing the two stars slightly closer together.

The special properties of binary pulsars allow this reduction to be measured directly. While this is a rare configuration of stars, such a system was first discovered by Russell Hulse and Joseph Taylor in 1974. The Hulse-Taylor binary has now been observed for over forty years, and the observations match perfectly with the predictions of general relativity. The system loses energy at precisely the rate it should if gravitational waves are given off at the level predicted by general relativity. For the discovery of this binary pulsar system, Hulse and Taylor were awarded the 1993 Nobel Prize for Physics.

Gravitational waves are the response of spacetime to a disturbance. How many different types of gravitational wave are there? In three spatial dimensions, there are precisely two types. This is – accidentally, it turns out – the same as the number of different light waves. There are also two types of light wave, distinguished by their polarisation. General light is a random admixture of both types of polarisation. By inserting filters that transmit only one type, the polarised components can be extracted separately. By inserting crossed filters that require first one and then the other type of polarisation, almost all light can be blocked, and this effect is used in sunglasses to reduce the glare and intensity of light.

Why does visible light have two polarisations? The answer is 'because there are three dimensions'. A light wave in two spatial dimensions would have only one kind of polarisation, and a light wave moving in four spatial dimensions would have three kinds of polarisation. In general, a light wave moving in D spatial dimensions has a total of $(D-1)$ possible ways that it can be polarised. The underlying reason for this is that a light wave is an oscillation in the electromagnetic field. It is a fact that the oscillation can occur in any direction – *except* along the direction the wave is travelling. For a wave moving in D spatial dimensions, there are exactly $(D-1)$ independent ways for the field to point, accounting precisely for the possible number of polarisation states.

The mathematics generates a similar but slightly different relationship for gravity waves. It is a fact that gravity waves require the specification of not one but two directions orthogonal to the line of travel. The mathematics then results in gravity waves in D spatial dimensions having $\frac{D(D-1)}{2} - 1$ possible 'polarisations'. In three spatial dimensions, this formula allows for two possible types of gravity wave.[2] This is the same as the number of possible light waves, although this appears to be coincidence and no deep reason is known for this similarity. It is worth noting that for two spatial dimensions, the above formula gives the answer of zero. The 'zero' is surprising, but correct. The gravitational force is actually not a dynamical force in less than three spatial dimensions, and gravitational waves are unable to propagate.

In a world with two spatial dimensions, there are no gravitational ripples. However as the dimensions increase the possibilities proliferate, and once we reach nine spatial dimensions, there are a total of thirty-five possible forms of gravitational wave.

Three spatial dimensions give two forms of gravitational waves. Nine spatial dimensions give thirty-five forms of gravitational waves. We started with the question: how does physics with nine spatial dimensions differ from physics with three spatial dimensions? These statements show that one partial answer is that with additional dimensions there are both additional forms of light wave and additional forms of gravitational wave. This represents a general truth about theories with additional dimensions. The more dimensions there are, the more ways there are for waves to propagate and the more types of polarisation there are.

What are these polarisation types and what do they correspond to? Polarised light waves correspond to oscillations of the electromagnetic field along directions transverse to the direction of motion. With extra spatial dimensions, the extra polarisation possibilities of light correspond to internal electromagnetic oscillations inside the extra dimensions. While there is still a wave transporting energy from A to B, the field oscillations now live in the extra dimensions.

The same is true of gravitational waves. If any extra dimensions exist, we cannot resolve them, and the gravitational oscillations are entirely confined within the extra dimensions. Whatever it might be on the smallest possible scales, the effective number of spatial dimensions is three on all distance scales we have measured. All our physics is described in terms of laws involving three spatial dimensions. How would extra-dimensional physics manifest itself if we are forced to express it in terms of laws that involve three, and only three, spatial dimensions? What has happened to all these extra gravitational polarisations?

The answer is that living in our world, we see these extra polarisations as new particles, all of whose interactions are of gravitational strength. Mathematically, this interaction strength follows from the origin of these particles in the fundamental theory as gravitational polarisation modes. The traces left

---

[2]For a wave propagating in the z-direction, the counting in three spatial dimensions is that there is an xx polarisation, an xy polarisation and a yy polarisation: except that there is a further technical condition, called the 'traceless' requirement, which enforces a relation between the xx and yy polarisations. This is responsible for the '-1' in the formula and results in two polarisations overall.

within our three-dimensional world by the extra dimensions are additional new particles, which interact by neither the electromagnetic, weak or strong forces. Instead, all their interactions are only by forces of gravitational strength.[3]

One effect of extra dimensions on the observable world is then through the existence of new, additional particles with gravitational-strength interactions, and only gravitational strength interactions. This feature is not the only effect of extra dimensions. It is, however, the most universal consequence of extra dimensions, which is the reason I focus on it here. To generate this feature, it does not matter how many extra dimensions there are, it does not matter how they are curved, it does not matter what their topology is, it does not matter whether they resemble a rainbow or a rugby ball, and it does not matter whether they are described by classical or quantum geometry. It is a universal feature. If extra dimensions are present, they invariably lead to the existence of new additional particles whose interactions are at levels vastly weaker than any of the particles present in the Standard Model of particle physics.

Such particles are called 'moduli'. The physics of moduli is the main topic of this chapter. I have so far given one route to the existence of moduli by viewing them as extra-dimensional polarisations. There is also another way to motivate the existence of moduli, and to understand their nature. This is to understand moduli as counting the number of ways one can change the extra-dimensional geometry without expending large amounts of energy. This draws on an analogy – an analogy that in the mathematics is in fact extremely close – between the physics of extra dimensions and the physics of oscillations of surfaces.

We know that different surfaces make different sounds because they vibrate in different ways. A drum is not the same as a bell, and a bell is not the same as a violin: the oscillations of each are *sui generis*, with each instrument having its own unique patterns of vibration. Furthermore, each instrument can also ring in many different ways – when a violin is bowed, its sound comes from vibrations at many different frequencies.

What in the name of Guarneri has this to do with moduli? The mathematics shows that you can view moduli as the oscillatory modes of the extra dimensions. This is not a poetic metaphor or some loose platitudinous analogy. It is as direct and as true a statement as one could hope to make without equations. The reason for this is that in Einstein's theory, geometry is dynamical. If you strike it, it rings. One 'strikes' geometry through a 'violent' nearby event: for example, through the gravitational collapse of a star to form a black hole. In theories with extra dimensions, a sufficiently violent hit will

---

[3]This paragraph is morally correct, but it comes at the cost of finagling the details. In particular, I have talked about moduli as having interactions of gravitational strength. However, these interactions are not precisely the same as those of the familiar gravitational force. Moduli additionally also interact with the familiar gravitational force, with a strength proportional to the mass. However the gravitational-strength interactions referred to here are different, and are really legacy interactions from the original ten-dimensional gravitational theory. These are comparable in strength, but different in form, to the standard four-dimensional gravitational force.

also cause the extra dimensions to vibrate. The energetics required for this to happen are large. Nonetheless, if the hit is large enough, the extra dimensions will vibrate. Just as for a violin, the manner of vibration will be as a superposition of many different oscillatory modes.

The mathematical statement is that every possible oscillatory mode of the extra dimensions appears in the lower-dimensional theory as a different particle. In a theory with extra dimensions, the types of particles reflect the structure and geometry of the extra dimensions. The superposition of many different frequencies of oscillation corresponds to the creation of many different types of particle. In this language moduli correspond to the deformations that are in a sense easiest to make – they cost the least amount of energy, and require the minimal amount of 'bang'.

These pages may have come across as hard to follow and mathematically heady. It is worth restating the key (and true) point. In theories with extra dimensions, one of the traces left by the extra dimensions in our three-dimensional world is the existence of additional particles whose interactions are exceedingly weak, and indeed only of gravitational strength. These particles are called moduli, and their existence is an unavoidable prediction of any theory including extra dimensions. If string theory is true, the existence of moduli is also a true statement about nature. Evidence for the existence of moduli would also be evidence for the existence of extra dimensions.

10.2   WHAT MAKES MODULI AND WHAT MODULI MAKE

This looks like big progress. Extra dimensions generically predict the existence of a new type of particle that has not yet been observed! The way to make the concept of extra dimensions scientific now appears clear. All we need to do is to devise an experiment that tells us whether moduli do or do not exist. If this experiment turns out positive, we have positive evidence for the reality of extra dimensions. If it fails, then we can falsify the existence of both moduli and extra dimensions. The history of particle physics is all about devising experiments to look for specific particles – we just need to think of an experiment to look for moduli.

Unfortunately, however appealing this sounds, it is difficult to do. It is extraordinarily hard to devise any laboratory-based experiment that can look for moduli. The reason for this comes from the most distinctive feature of moduli: all their interactions are of gravitational strength, and gravitational interactions are by far the weakest. This fact has been discussed before, but it is worth re-stating. As I write these words, my puny humanoid hand and its puny humanoid muscles can lift the pen up against the gravitational pull of the *entire earth*. There are almost as many kilograms in the earth as there are atoms in a kilogram. The number of protons and neutrons in the earth is so large that it is more than the number of grains of sand that could fit inside the volume of Pluto's orbit around the sun – and yet the combined gravitational pull of every single one of these particles, added together, is insufficient to restrain the pen against the electromagnetic tug generated by my muscles. If this is what so many known particles can fail to do via gravitational interactions, what hope do we have of making an experiment to detect a single unknown particle whose interactions are of similar strength?

While the interactions of moduli are extremely weak, they are not entirely zero. What do these new interactions do? One thing they do is to provide ways for heavy moduli to decay to light particles. Generally, heavy particles always tend to decay into light particles if they can. The Higgs boson decays to bottom quarks. The bottom quark decays to other, lighter quarks. The muon decays to an electron and a muon neutrino. Heavy particles are unstable – they live for a short period of time before falling apart into lighter and more stable daughter particles. The series of decays as heavier particles decay to less heavy ones, which again decay, which again decay, generates a particle decay chain. In these decay chains, weight loss is rapid and irreversible. Once particle decay has occurred, it cannot be undone except by re-colliding particles at high energies using accelerators.

A decay is one form of interaction, and the speed of decay depends on particle mass and interaction strength. Heavier particles decay quicker, and more strongly interacting particles decay quicker. While a particle decaying via the strong force has a typical lifetime of one million-billion-billionth of a second, one decaying through the weak force lives much longer, and in some special cases the lifetime can reach human timescales. A free neutron lives on average for fifteen minutes before decaying via the weak force to a proton, an electron and an anti-neutrino. This is atypically long – a more normal value for a weak interaction lifetime is provided by the decay of the muon, with a lifetime of around a microsecond.

The consequence of all this is that if a particle only barely interacts, it can survive for exceedingly long times. Such particles are very hard to create due to their almost non-existent interactions. However, once created, they are also almost impossible to dispose of. As interactions of gravitational strength are the weakest of all, particles that only interact in this way have the longest lifetimes of all. Once they have, with great difficulty, been created, there is nothing one can do other than wait for them to decay.

Objects that interact weakly live a long time and objects that interact strongly do not: this fact is not confined to particle physics. The safest way to store fragile *objets d'art*, whether they be jewellery, tapestries or manuscripts, is by hiding them away so that they lie unjostled and untouched by dirty hands and bacteria-laden breathing. Many a librarian responsible for ancient books fears nothing more than the books being read, and grubby fingers poring over them – hence the rush to digitise ancient manuscripts.

To look for observational consequences of moduli though, we do need a way of making them. The harder it is to make a particle, the more extreme the conditions that are required for an efficient particle factory.

As we saw in chapter 3, one of the great truths about the early universe is that it was both extremely hot and extremely dense. Our universe at its start was a more rambunctious and callithumpian place than at any time subsequently. It was also more energetic than even the most fiery locations, such as the interior of supernovae, present in the universe today. The gargantuan amounts of energy present in the early universe allowed for the production of particles then that could never be produced at more irenic later epochs.

In particular, this includes moduli – which despite their exceedingly weak interactions could be made when the universe was very young.[4]

Moduli are hard to make, and hard to get rid of. Once moduli have been created, they survive for what is, by the standards of the early universe, an aeon. Aeons are relative, as here even a microsecond can count as a long time. Timescales in the early universe are measured, without blushing, in picoseconds or nanoseconds. As the universe moves through these ages, it grows. The universe at an age of a nanosecond is by volume over a million times larger than it was at the age of a picosecond.

Moduli live for a long time – what does this mean? This is best understood through ratios. What was the universe like when the moduli were made? What was it like when the moduli decayed? The quantity to compare is the age of the universe at each epoch, and the ratio between these times can be enormous, easily reaching a factor of ten followed by twenty-four zeroes. This ratio in ages is the same as that between the universe today and the universe at only one microsecond old. At the epoch of moduli decay, the universe is vastly older than at the epoch of moduli creation. Such decaying moduli are coelacanths of the early universe, surviving relics of a distant past.

There is another important aspect of moduli in the early universe. Their Methuselan properties also imply that, by the time moduli eventually decay, they are expected to dominate the energy density of the universe. This last sentence is another way of saying that if you look at what makes up the universe at the time at which the moduli decay, it is expected to consist of moduli, moduli, and moduli – and almost nothing else. As moduli linger while all else decays, the proportion of the energy of the universe that is in the form of moduli is expected to grow and grow, until it is almost one hundred percent by the time that moduli finally decay.[5]

In this respect, there is another parallel between the behaviour of moduli and the behaviour of a struck bell. When a bell is struck, the timbre of the sound changes with time. The higher-pitched notes die out more rapidly, while the deep low frequency notes last longer. While the initial strike of the clapper

---

[4]A technical footnote: the production of moduli occurs by a process called 'misalignment'. The large energy of the early universe drives it far away from the cold and stable state it is in now. This 'misalignment' of the state of the early universe compared to its eventual form in the late universe leads to large numbers of moduli being produced as the universe relaxes and settles down into its late-time state.

[5]The more technical justification for this is that earlier decaying particles produce relativistic daughter particles, and in an expanding universe the energy in relativistic particles rapidly dissipates due to the Doppler effect. This leaves only the energy in the non-relativistic particles that have yet to decay – the moduli.

generates a superposition of notes of many frequencies, a couple of seconds later only the most long-lasting low frequency notes can still be heard.

On this analogy, particles with strong interactions correspond to the high-pitched notes, while the moduli correspond to the deep bass frequencies. As with the deep bass notes of the bell, in the early universe particles that interact strongly soon decay and disappear, while the particles with extremely weak interactions are able to survive for a long time. The universe then goes through a phase where it is dominated by the longest lasting particles – the moduli.

The summary of the above is that moduli were produced in the extraordinarily early universe, before disappearing through decays to other particles in the merely very early universe. In between – they dominated. Despite this, our first thought is that this does not matter very much. The time at which moduli particles decayed and disappeared was long before there were atoms, let alone stars or galaxies or people. Like the dinosaurs, moduli had an epoch of importance, but have now all disappeared.

Of course, we actually know rather a lot about dinosaurs. As with the dinosaurs, we can still look for fossil traces of the existence of moduli. The time at which moduli decayed is also late enough that what happened then can be connected onto the physics that we do observe. The equations that governed the universe at this period are equations that we understand. They do not require quantum gravity. They do not require extra dimensions. They do not require spacetime foam. They are the standard and familiar equations of particle physics and general relativity that have been tested repeatedly against experiment, and that we know are correct. If something happened then, we know how to evolve it forward to the present day.

Let us pause for breath. I have argued above that moduli are important because we expect the universe to go through an epoch where it is filled with moduli and only moduli. The moduli are gone now, but for a time they were everything, and we can hope to find legacies from this era. The history of the universe subsequent to this point is determined by what actually happens when the moduli decay, and it is to this we now turn our attention.

What happens when a modulus decays? In a decay, the mass-energy of the modulus particle is converted into the energy of the daughter particles produced. The heavier the modulus particle, the more energetic the daughter particles are. Suppose a modulus decayed into two photons, so that the decay results in two photons heading out in opposite directions. As the mass of the modulus particle increases, these photons appear respectively as radio waves, visible light, X-rays or gamma rays.

For our purposes, there are two distinct possible ways for the decay to occur. The modulus can either decay to particles from the visible universe, or it can decay to particles from the invisible universe. The invisible universe is the dark universe, which is by definition dark: it consists of those parts of the universe that we cannot see, and which do not easily interact with photons.

What makes up the dark universe? First, it includes dark matter – stuff moving at speeds far lower than that of light, and that is heavy and can be weighed. As discussed in chapter 4, we know that dark matter is present, but we have little firm idea what it actually is. The quest to identify dark matter is one of the largest areas of research in particle physics and cosmology.

The dark universe also includes dark energy. This represents the energy of space itself. It is present in even the emptiest vacuum, and its magnitude is proportional purely to the volume of space. Dark energy is also hard to understand, but for a different reason. With dark energy, we have a good idea *what* it is, and observations are entirely consistent with this idea. Our problem is that we have no understanding of *why* it is there. Quantitative attempts to calculate the amount of dark energy give values that overshoot the data by a factor of at least ten to the power of sixty, and no good resolution of this problem has been found.

My focus here is on an additional possible element of the dark universe. This is called *dark radiation*, and it refers to stuff that is both dark and relativistic, moving at or extremely close to the speed of light. The word 'radiation' is used in analogy to light. Conventional radiation consists of a massless particle, the photon, moving at the speed of light. The visible universe is full of photons, which we could call 'visible' radiation. Dark radiation is the analogue, except with invisible particles.

This is not actually so exotic. We know that at least one particle in the universe, the photon, travels at the speed of light. Once you accept the existence of dark matter, you accept that there are still unknown aspects to the universe. The history of particle physics, with its repeated discovery of new particles, should render the wise sceptical of any suggestion that all the particles in the universe have now been discovered. The notion that the dark universe contains both relativistic and non-relativistic particles should not be shocking, as it would just mirror the properties of the visible universe.

If dark radiation exists, how would we ever know? First, relativistic energy density – 'radiation' – behaves differently from non-relativistic energy density – 'matter'. For a fixed total amount of energy, the universe expands more rapidly if that energy is in the form of matter than if it is in the form of radiation. This statement is not at all obvious, but its truth follows from Einstein's equations of general relativity.

In an expanding universe, the density of energy in matter also dilutes more slowly than the density of energy in radiation. This can be understood by imagining a fixed number of particles, either relativistic or non-relativistic, within a box whose volume grows as the space inside it is stretched out. As the box grows, the number of particles per unit volume decreases. The volume of the universe is bigger while the total number of particles remains the same. In both cases, the number density of particles per unit volume decreases in exact proportion to the increase in volume.

The mass, and therefore the energy, of non-relativistic particles remains the same as the universe expands. However, as the volume increases the energies of relativistic particles decrease. In an expanding universe, every part of the universe is running away from every other part. In the same way that an ambulance siren decreases in pitch as an ambulance accelerates away from you, the energies of relativistic particles decrease as the universe expands – and so the overall fraction of energy in radiation decreases.

As the universe expands and ages, matter becomes more and more important and radiation less and less important. This is the same logic that explains why moduli would have come to dominate the universe. Moduli are long-lived matter, and so they win out over any radiation that was present.

It follows that the effects of radiation become ever more important as one goes back in time. To look for the effects of dark radiation, we need to look at the young universe. As noted in chapter 3, there is one visible relic from this era. This is the cosmic microwave background, the diffuse glow of microwave light left over from the early universe. If dark radiation were present, the detailed properties of the cosmic microwave background are changed slightly. For a fixed amount of energy pie, if some is given to additional invisible radiation, there is less of the pie left for visible radiation. This leads to small modifications in the rate at which the universe expands, leading to small modifications in the properties of the microwave background.[6]

The theory is clear. We know what the implications of dark radiation are. Current experiments are not, however, quite sensitive enough to tell us whether dark radiation is present or not. Hints have come, grown and shrunk. The quality of experimental precision is however improving rapidly, and a decisive verdict is expected over the next decade.

Let us recall why we introduced dark radiation. We have argued that, if moduli exist, the universe will go through an epoch where its energy is almost entirely in the form of massive moduli particles. The next step in the story of the universe occurs when the moduli decay. As we shall soon see, these decays can lead to dark radiation.

When moduli decay, their mass-energy is given to daughter particles. As we have seen, there are two basic possibilities. Part of the time, the moduli will decay into the particles of the Standard Model: the electron, the quarks, the W bosons, the Higgs boson and others. When this happens, all the mass-energy of the modulus gets turned into a hot and energetic plasma of quarks, gluons and photons, as if in a superheated oven. Gluons scatter off quarks which emit W bosons which decay into quarks which scatter off quarks ... the energy of the original particle is dissipated across hundreds and thousands of particles scattering off one another and exchanging their energy to form a hot thermal bath. This hot thermal bath is the Hot Big Bang. As the universe expands, this hot bath gradually cools down, until it becomes today the afterglow of the Big Bang that is the cosmic microwave background.

---

[6]To be precise, this results in the spectrum of the microwave background, when expanded in spherical harmonics on the sky, having less power on small scales than would otherwise be expected.

While this is the light side of the decays, our interest is in the dark side. I introduced moduli as extra-dimensional modes of gravity and a necessary consequence of a fundamental theory with more spatial dimensions than three. Gravity is universal. It loves everyone and everything, and no form of energy is exempt from it. The consequence of this is that while moduli will decay into the particles of the Standard Model, they will also, part of the time, decay into the dark sector where interactions with all familiar particles are exceedingly weak.

The democracy of gravitational interactions implies that moduli interact with the hidden sector just as strongly, or weakly, as they do with the visible sector. Moduli decay to hidden particles at a similar rate to their decays to Standard Model particles, and in decaying to the hidden sector they can produce dark radiation.

Our focus here is on these invisible decays to the ghosts of the dark sector, the light particles with ephemeral interactions with our visible world. Energy released into these particles does not heat the universe. Instead, such particles continue through the universe as they were formed, travelling at the speed of light but neither scattering nor interacting.

Are these particles neutrinos? Neutrinos, famously, barely interact with matter. They are the most feebly interacting of all known particles, and they can pass through thousands of kilometres of solid rock as if it were empty space. Indeed, experimental searches for dark radiation are often loosely called searches for additional neutrino species. This reflects the fact that at the time the cosmic microwave background was formed, when the universe was four hundred thousand years old, neutrinos are dark, relativistic particles to which the universe is transparent. Neutrinos behave precisely in the same way that dark radiation does, and so when searching for the effects of additional dark relativistic particles in the cosmic microwave background, it is necessary first to subtract the known contribution from neutrinos.

Despite all this, in the very beginning the universe was not transparent to neutrinos. When less than one second old, the universe was so dense that it was opaque to neutrinos. During this period, neutrinos were like any other particle. When a neutrino was produced, it would interact with the thermal bath and deposit its energy back into this hot soup. This is because neutrino interactions, while weak and tenuous, are still vastly stronger than purely gravitational interactions. While a neutron star is dense enough to stop neutrinos, a 'gravitational neutrino' would pass straight through such a star in the way neutrinos pass straight through the earth.

In the next section I will discuss one specific candidate particle for 'gravitational neutrinos'. The broader general feature, however, is that the universe is transparent to such particles from the moment of their birth. Once they are produced by decaying moduli, they travel freely throughout the universe, passing through whatever they encounter on the way. The fact that the interaction strength is so weak means that as the universe expands such particles continue to travel through the universe, even today, at the speed of light. They are the Flying Dutchmen of the universe; unable to stop and unable to interact, they are condemned to travel as dark radiation permanently and forever.

Why is this interesting? From a theorist's perspective, what is striking about dark radiation is that, given relatively few assumptions, its existence is very hard to avoid. This comes from its origin in a moduli-dominated epoch of the universe, when the energy density of the universe lay almost entirely in the form of moduli. These moduli are bridges between the visible sector of Standard Model particles and the gravitational world of ephemeral weakly coupled massless particles. Moduli at the time of their decay are like a pencil balanced on its tip at the edge of a chasm – sometimes the pencil lands safely, and sometimes it disappears into darkness. The mathematics of quantum mechanics guarantees that, given enough moduli, some will always end up decaying to the invisible shadow world of dark radiation.

Calculation shows that this process should lead to observable levels of dark radiation. The origin of the moduli as gravitational, extra-dimensional modes makes their decays democratic. Gravity talks alike to big and small, and is the most democratic of all forces. This gives the exciting possibility that if experiments over the next decade achieve their claimed sensitivity, the existence of dark radiation may be established.

Dark radiation represents one way that apparently inaccessible physics – moduli particles with gravitational-strength interactions that ceased to be present when the universe was a microsecond old – can leave observational traces that are measurable today using the technology of today.

It is a cosmological analogue of a dinosaur footprint. The argument involves a chain of inferences, each of which is more than reasonable by itself. The conclusion involves a quantity whose existence can be determined, and where the experimental precision on the measurement will only improve. Over the next few years, observations from a variety of either balloon-based or ground-based telescopes may determine the existence of dark radiation. These telescopes observe the sky from the rarefied air of the Atacama desert in Chile – as for the Atatcama Cosmology Telescope – or from the pristine environment of the south Pole, as for the South Pole Telescope or the Spider balloon.

If the existence of dark radiation is established, it will be a great triumph for physics. It would be a discovery equal to the realisation that dark matter exists. As with dark matter, it would reveal the existence of a new, unknown, contribution to the energy budget of the universe.

Of course, it may not happen. The arguments here involve reasonable assumptions, but they are not guaranteed to be correct. The scientific landscape is littered with the desiccated corpses of ideas that once seemed appealing but could not command observational support. That is the way of the subject, and the normal operation of normal science.

However, suppose dark radiation is found to exist. Can one do better? With dark matter, we know it is there as we can weigh its total effect, but we do not know *what* it is. So far, every attempt to detect dark matter directly has failed. While there are many candidate ideas, we do not know whether any are correct, and we have no positive experimental evidence to guide us.

If it exists, a similar problem would occur for dark radiation. In everything I have just said, it was crucial that dark radiation particles could survive in the dense and crowded environment of the early universe. Particles that are produced at sub-microsecond times can only survive for fifteen billion years by not interacting. The same quality that ensures their survival makes them hard to detect – one cannot be simultaneously both a social butterfly and a ninja. The feebleness of their interactions allows them to survive to the present day, but also renders it difficult for us to know what they are.

While we may be able to measure the total amount of dark radiation through weighing its overall gravitational effect on the expansion of the universe, it is hard for us to find out more. How can we effectively ask the question: who are you, and from where do you come?

For one special type of particle, called either an *axion* or an *axion-like particle*, this problem may be evaded. While this type of particle has never yet been observed, it arises frequently in string theory.

10.3   AXION-LIKE PARTICLES

The axion is the first, and so far only, particle to be named after a brand of washing-up liquid.[7] As mentioned in chapter 4, it originates as a proposed solution to a specific problem: why does the charge distribution of the neutron not have a preferred direction? The neutron, a bound combination of two down quarks and one up quark, is a neutral particle, and so can be viewed as smeared-out positive charge combined with smeared-out negative charge. To the limits of experimental accuracy, however, the centroid of the positive charge distribution and the centroid of the negative charge distribution are in exactly the same location. There is no weight of positive charge on one side balanced by negative charge on the other.

This can only happen if one particular parameter – the so-called theta angle – in the equations of the strong nuclear force is equal to zero to within around one part in ten billion. This may not be impossible, but it does not conform to good taste.

There is a simple and elegant solution to this problem. This simple and elegant solution was found by Robert Peccei and Helen Quinn in 1977 and is now called the Peccei-Quinn mechanism. In this solution, the theta angle is not regarded as a fixed constant that the Prime Mover initialised when booting up the universe. Instead, theta is treated as a dynamical entity that is free to change its value according to the prevailing conditions. Subtle effects of the strong nuclear force lead to a preferred value for theta – which is

---

[7]Historians of science can use the names of scientific concepts as a sociological history of the discipline, by tracing the fluctuating linguistic dominance of Greek, Arabic, Latin, French, German and English. What future historians will make of 'axion' is best left to posterity.

precisely zero.[8] Whatever the initial value of theta, it always ends up at zero, precisely where observation requires. This turns a problem of fine-tuning into a problem analogous to why a compass needle always points north: it aligns itself with a field.

Even after theta has relaxed to its equilibrium value of zero, it still has residual quantum mechanical fluctuations. Soon after the publication of Peccei and Quinn's original paper, Steven Weinberg and Frank Wilczek showed independently that these residual fluctuations were mathematically equivalent to the existence of a new particle – the axion. The particle was called the axion because, like its ablutionary namesake, it washed away a problem, in this case the problem of the charge distribution of the neutron.

Weinberg and Wilczek demonstrated certain properties of the axion. They showed that the axion is always light and that the axion always interacts weakly. Experiments were quickly devised to look for axions that were light and interacted weakly, but with no success. If the axion exists, it is not just light and weak, but extremely light and extremely weak. As it became apparent in the 1980s that the axion, if it existed, was invisible at any conventional experiment, the expression 'invisible axion' was coined for such particles.

Since then searches for the 'invisible axion' have continued, using dedicated experiments and novel techniques that are radically different from banging particles together at high energies. While no successful detection has yet occurred,[9] new experiments such as the International Axion Observatory continue to be built, and discovery may yet occur.

From the perspective of a particle theorist, there are three defining features to the axion. The first feature is that the axion is light. The second feature is that the axion is very weakly coupled. The third and final feature is that the axion has a special interaction with the strong force, which enables it to solve the problem of the charge distribution of the neutron.

There is a generalisation of this to consider particles with only the first two of these features. These particles are also extremely light – for the same reasons that the axion is light. They are also weakly coupled – for the same reasons that the axion is weakly coupled. However, they lack any relation to the strong nuclear force. Such particles are, for hopefully clear reasons, known as axion-like particles, while the definitive article associated to *the* axion is reserved for the particle that couples to the strong nuclear force.

I now want to explain briefly *why* axions and axion-like particles are so light, although this more technical discussion can be skipped if desired. The mass for the axion is not larger than one billionth of the mass of the electron, and masses for axion-like particles can be far lighter still. What could

---

[8]In more technical language, the strong force generates a potential for theta, and the minimum of that potential is where theta vanishes. Even more technically, this potential arises from non-perturbative 'instanton' effects.

[9]Which is a shame for Helen Quinn, who would become, if the axion were discovered, the first female Physics Nobel Laureate since Maria Goeppert-Mayer in 1963.

keep these particles so light, indeed so much lighter than almost any other known particle?

The danger here is that quantum fluctuations in general give large masses to particles. There needs to be a special and powerful reason to keep a particle a flyweight in a world of cruiserweights. This special reason comes from one of the most powerful principles in physics, introduced in chapter 3: the principle of symmetry. There is a symmetry in the equations for axionic particles such that, if it were exact, the axion mass would be exactly zero.[10] Roughly, the mass of a particle tells you the energy that is associated to quantum fluctuations of that particle – the larger the mass, the greater the energy for any given fluctuation. The symmetry implies that the energy of a fluctuation is independent of its size: this is only true if the particle is massless.

If this symmetry were flawless, the axion would be perfectly massless. However this symmetry is not perfect and it has minor blemishes. Whilst it holds for almost all calculations, it is violated by a certain class of small effects ('instanton' effects). These Lilliputian, exponentially small effects are the only way for axions or axion-like particles to obtain a mass, forcing this mass to be very small in magnitude.

The tiny masses of axion-like particles are one reason to be interested in them. As they are so light, their production is not obstructed by energetic requirements. They offer an attractive target to those who balk at the scale and cost of arranging thousands of superconducting magnets in underground tunnels.

However, another reason to care about axion-like particles is that they arise ubiquitously in string theory. If string theory is a true theory of nature, axion-like particles should exist. Why?

Our discussion of moduli emphasised the fact that within string theory, the observed particles in four dimensions are geometric guano from ten dimensions. Like pigeon poo from above the clouds, they are visible droppings from an extra-dimensional geometry that we cannot directly access.

Axion-like particles share this origin: they are the four-dimensional legacies of a particular feature within the geometry of the additional six spatial dimensions. What is this feature? It is the presence of subspaces within this geometry that one cannot contract away to a point.

How can you visualise these? One way is by imagining stretched spandex under tension. Its tension makes it want to collapse to a point, and if unconstrained it will shrink like a deflated balloon. However, if there is a surface it is wrapped around – imagine a bald man's head in a morphsuit – it cannot collapse. Instead it ends up wrapping the surface and maintaining a finite area.

The mathematical statement is that whenever such a non-contractible surface or hypersurface appears within the extra-dimensional geometry, an

---

[10]Technically, this is a shift symmetry that states that the value of the axion potential is independent of the value of the axion field.

axion-like particle exists in the four-dimensional theory. In fact such surfaces are common, and particular geometries may even include several hundred of them, which would lead to several hundred different axion-like particles in our four-dimensional world – all very weakly interacting.

I want to summarise the key points of the last few pages. Axion-like particles are naturally extremely light and with extremely weak interactions. While difficult to see because of these weak interactions, they are far lighter than more familiar particles, and so there is no energetic obstruction to producing them. Axion-like particles also occur generally in string theory. If string theory is indeed a true account of nature, such particles should exist.

The relevance of these particles is that their properties make them ideal candidates to be dark radiation. As they are so light, they behave as relativistic particles and count as radiation. As they interact so weakly, they are dark and able to propagate freely through the universe without returning their energy to the visible sector. If axion-like particles are present, moduli are expected to decay to them at a similar rate to their decays to the particles of the Standard Model. Axion-like particles are realisations of our previous generalities: they represent actual candidate particles to be 'gravitational neutrinos'.

10.4   COSMIC MAGNETISM

Axion-like particles are attractive candidates to be dark radiation. So what? None of this helps if such particles are permanently dark. Experimental verification is at the heart of science. Invisible fire-breathing dragons have the same status as invisible white unicorns, and invisible axions may have the same status as invisible justmadeupons.

As the universe expands and becomes less dense, most dark particles become harder and harder to see. Sight requires photons, and in most cases photons require interaction. Particles need to find other particles in order to interact and become visible. The fewer particles around, the smaller the chance of any interaction occurring.

What is also interesting about axion-like particles is that they are an important exception to this rule. While the visibility of most types of dark particle falls away with the growing age of the universe, for axion-like particles the visibility can, under particular circumstances, increase. This occurs because axion-like particles have the ability, in the right environment, to turn into photons. It is clear that if this happens they become instantly visible: photons are the quantum building blocks of light, and the means through which we see the universe.

What is the right environment? It means here a sufficiently large magnetic field stretching over a sufficiently large distance. Provided the magnetic field points at right angles to the direction of travel of the axion-like particle, there is a small but finite probability for the axion-like particle to convert into a photon. While I assert this as a true statement, I will not justify it. There is no simple way to explain it. After several graduate-level courses in quantum field theory, it becomes 'easy' to compute the conversion probability. This probability grows with both the square of the magnetic field and the square of the length over which the magnetic field extends. To make conversion more likely, you need to embiggen either the size of the magnetic field or the length over which it stretches.

This principle is already used by several experiments that search for axions. For example, at CERN there is an experiment called the CERN Axion Solar Telescope. It uses a large superconducting test magnet left over from the Large Hadron Collider, pointing the magnet at the sun. It is like a conventional telescope with the roof shut. The experiment is inside a closed building, and slowly rotates the magnet so that it is always tracking the position on the roof that matches that of the sun in the sky. The experiment searches for axions that may be produced in the deep interior of the sun. Such axions would stream freely through space to us, through the roof of the building, and then reconvert to photons within the interior of the magnet. LHC magnets are used to maximise the signal. These have large magnetic fields and, at ten metres, are also relatively long.

Large magnetic fields help because conversion grows with the square of the magnetic field. However conversion also grows with the square of the field's extent. If axion-like particles form part of dark radiation, they have streamed through deep space during the entire history of the universe. As the universe has aged and expanded, it has made structures that are elongated over enormous scales. The best known are galaxies, beautiful patterns etched on the sky and stretching over distances greater than one hundred thousand light years. There are still larger scales though. Galaxies have come together into galaxy clusters, assemblies of hundreds or thousands galaxies orbiting one other on distances of several million light years.

Both these gargantuan objects contain ordered magnetic fields. Within our own Milky Way galaxy, the magnetic field approximately traces the pattern and orientation of the spiral arms. This magnetic field is a large-scale structure within the Milky Way, continuing to point in the same direction across thousands of light years. The same is true of galaxy clusters. Although galaxy clusters lack the beautiful ordered spirals that can occur in galaxies, they still contain large-scale magnetic fields, which retain their orientation even across distances as large as a hundred thousand light years.

It is true that the magnetic fields in galaxies or clusters of galaxies are not large. They are over a million times smaller than the magnetic field found on the bumpers of a child's toy train. However, a hundred thousand light years is a jolly large distance, and conversion depends not just on this distance but on its square. While the magnetic field of a cluster of galaxies is smaller than that of the CERN Axion Solar Telescope by a factor of around ten billion, it extends over a length larger by a factor of ten billion billion. Doing the mathematics, one finds that a cluster of galaxies is more efficient than the CERN Axion Solar Telescope at converting axions to photons by a factor of ten followed by eighteen zeroes.

Some of the best axion convertors in the universe, then, have been provided by nature for free. The making of galaxies or galaxy clusters cost us nothing. It took large amounts of time and energy, but it happened a long time ago, and we did not need to wait for the making. The finished products of nature's bounty are present in the sky and available for us to look at – we just need a telescope.

Of all places in the universe, clusters of galaxies are most efficient at converting invisible axion-like particles into visible photons. They maximise the product of size and extent of magnetic field. If dark radiation exists in the form of axion-like particles streaming through the universe, as these particles pass through a cluster of galaxies they will convert and produce an excess of photons, correlated with regions of maximal magnetic field.

It is this behaviour within a magnetic field that makes axion-like particles special – it is a distinctive property that applies to these particles and these particles alone. It is this behaviour within a magnetic field that makes axion-like particles predictive candidates for dark radiation. Given enough knowledge about the form of the magnetic field within galaxy clusters, it is possible to predict the structure and number of the photons that are produced. The determination of the magnetic field within a cluster of galaxies is also not exotic physics. It can be determined, albeit with errors and uncertainties, using other, known techniques.

What are the energies of the photons that are produced? These come from the initial energies of the axion-like particles. The exact calculations are not for here, but the most important fact about these is that the energy of each individual axion-like particle must be much larger than the energy of each photon that makes up the cosmic microwave background. The reason for this comes from what happened at the time the axion-like particles were formed.

As we have seen, these axion-like particles originate in an epoch of the universe where it was dominated by moduli, moduli and more moduli. The decay of a modulus to any particle of the Standard Model launches a high-energy Standard Model particle into the hot dense thermal plasma of the early universe. This is analogous to launching a highly energetic particle at the sun and asking what happens to it. It loses its energy. It bumps into other particles, which bump into other particles, which bump into more particles, and rapidly the energy of the original particle is shared out among thousands more. When an axion-like particle is formed, it loses none of its initial energy. It passes through the hot thermal plasma of the early universe like a ghost, giving its energy to none. Unlike particles of the Standard Model, it retains its original energy without loss.

While far fewer in number than the particles of the Standard Model, the energy of each axion-like particle would be far greater than the energy of the photons that make up the cosmic microwave background. In fact, for the best-guess parameters that one gets from string theory models of the early universe, the energy of each particle is around a million times bigger than the microwave photons that make up the cosmic microwave background. Instead of microwave energies, they have X-ray energies. If they convert to photons, they produce photons with energies at the level of the weakest X-rays.

We have said that the best place for converting axion-like particles into photons is within large clusters of galaxies. It turns out that there is indeed actually an excess in the number of soft X-ray photons coming from large clusters of galaxies. While clusters of galaxies are suffused with a hot gas that emits copiously in X-rays, at the lowest energies there appears to be more X-ray emission present than can be accounted for by this hot gas: there is a soft excess in the X-ray emission. This was discovered by Richard Lieu of the University of Alabama in the middle of the 1990s. Observations of a number of galaxy clusters, for example the relatively bright and nearby Coma cluster of galaxies, show a pronounced excess in the softest forms of X-ray emission. This excess emission has been sought for in a total of thirty-eight clusters of galaxies, and it is found in around a third of them.

This excess is at the right energy to come from a cosmic background of axion-like particles. It also occurs precisely from galaxy clusters, objects that have the largest magnetic fields extended over the largest regions. Could it actually be a sign of dark radiation made visible?

Possibly it is – but while not impossible, the most reasonable answer is 'probably not'. At this point it is important to state all the caveats on the set of ideas described here. Almost every attempt at finding new physics beyond the Standard Model fails. Most anomalies go away, even the ones produced by excellent and careful experimentalists. A failure to come up with an explanation for a phenomenon is not the same as the absence of an explanation for this phenomenon. Given any choice between an anomaly being explained by either new physics or by incompletely understood aspects of existing physics, the right choice is almost always the latter. Furthermore, the science of hot astronomical gases in astrophysical environments – sometimes known as gas-trophysics – is messy, and it is susceptible to both incorrect modelling and incomplete understanding.

For this reason, astrophysicists are justly nervous of any claims for new physics lying beyond the known laws of nature. For this reason, the evidence for dark matter was dismissed until it became so large and so overwhelming that there was no choice but to accept it. Most anomalies go away. The only way to move forward with anomalies is to take them and batter them with more data until they either grow or vanish.

For the case of the excess in soft X-rays and a possible relation to a pervasive stream of dark axion-like particles, the method for assault and battery is clear. In this scenario, the strength of the signal depends on the distribution of the magnetic field. The magnetic field is a property of the universe. It can be measured for different galaxy clusters using known techniques, and these techniques are improving. Given these measurements, predictions can be made for the amount of excess X-ray emission that is expected. Through the use of ever more precise X-ray telescopes, the existence and magnitude of the excess emission can be measured and compared to the prediction.

Either it will agree or it will not: and on this verdict the correctness of the scenario ultimately stands.

10.5   EXPERIMENTAL ENNUI

I have devoted considerable space in this chapter to describing one example of the use of string theory to generate testable ideas for new physics. This idea was born amid the physics of ten dimensions, but it lives in the world of observations. It might be right and it might be wrong. Like all ideas for new physics, it is probably wrong.

The reasons for devoting so much space to this particular idea are twofold. The first is unashamedly partisan – it involves my own work. I know these ideas, and their internal logic, well. They come from my routine professional life of writing papers, and the research of myself and my collaborators: Michele Cicoli, David Marsh and Fernando Quevedo.

However, the deeper and more important reason is to put sinewed flesh on the rather abstract bones of the notion that extra-dimensional physics at minuscule scales can leave traces within the majuscule world of physics accessible to observation. I have sought to provide one example of how the abstract concept of extra dimensions can lead to a set of ideas that have observational content and are clearly empirically testable. Furthermore, one would not stumble upon these ideas without thinking about the physics of extra dimensions.

For some – like myself – this is the principal motivation for working on string theory. It is understanding *this* universe that makes my heart beat quicker and the hairs on my arms stand up, and gets me eagerly hurrying to the railway station every morning on the way to work. I want to know something new about our world, and I want to understand the parts of it that are not understood. I respect those who seek the quantum-mechanical behaviour of gravity on the smallest scales, but this task does not enthrall me, and I equally feel little passion for the intrinsic mathematics of extra-dimensional geometry topped up with large doses of supersymmetry. My interest in string theory is in what it can offer to physics that can be probed by experiment.

This view is far from universal. It may seem odd, but most of those who work on string theory are essentially uninterested in any connections with experiment, any public claims they may make to the contrary notwithstanding. This is illustrated through a notion economists call 'revealed preference': to see what people really care about, look at what they do and not what they say. The largest annual conference in string theory is the Strings series, which in its heyday would attract up to five hundred participants. While the conference has a notional mandate to cover the whole of the subject, every year there are few if any talks that concern the physics of this world.[11] The physics

---

[11] Based on the talk titles at the Strings conferences in 2012, 2013 and 2014, my count is that barely one talk in ten, even on a generous interpretation, could be put in this category.

of ideal mathematical worlds is covered in plenty, but the physics of this world gets only minimal attention.

Along these lines, there are three kinds of objection made concerning connections to experiment. One of these objections deserves a small and limited amount of sympathy, while two deserve none at all.

The first objection is that there is no unique and unbreakable logical path that runs from string theory and the existence of extra dimensions to any observational proposal. Assumptions have to be made. The path involves qualifiers such as 'generic', as well as steps that may be reasonable but are not logically unavoidable. The thrust of this objection is that you end up with a set of ideas that may or may not be interesting, but they do not tell you anything about string theory or extra dimensions as there is no unique logical thread leading from start to finish.

The reason that this objection deserves some sympathy is that within the scientific ideal as presented in textbooks, ideas have exact consequences. These consequences can be tested experimentally, and a single well-chosen and well-conducted experiment is sufficient to decide whether an idea is correct or not. In the world of Pangloss, the idea 'there exists extra dimensions' would lead to a set of necessary consequences that are directly testable. If these consequences did not manifest themselves experimentally, then the idea would be falsified, and we would know that extra dimensions do not exist. Situations when this occurs – and it sometimes does! – are the scientific ideal: clean ideas cleanly tested. One example involved the Higgs boson at the Large Hadron Collider – there was a guarantee in advance that either the Higgs boson would be discovered, or the Standard Model of particle physics would be wrong.

The reason that this objection only deserves limited sympathy is that on the whole the world belongs to Candide and not to Pangloss. Science is messy. The mood of the route from theory to experiment is generally the optative rather than the imperative. Life is uncertain. Most of the time, the passage from theory to experiment has to involve qualifiers, hopes, assumptions and 'if this were also true' conditionals.

The second objection, which deserves no sympathy at all, is that most of the time the testable scenarios that are generated by the extra-dimensional muse fail. This deserves zero sympathy because this situation is the default situation when searching for new physical laws beyond those we already know. The search for new phenomena or new laws of physics is a long succession of failures punctuated by occasional and unpredictable success. Anyone contemplating a career looking for deviations from the established laws of physics must be prepared for hard work followed by failure. This is how the subject works; this is normal science; this is where success ultimately comes from.

This objection of probable failure is like a complaint from the armchair expert to the explorer that there is no point searching for unicorns, as

---

The ratio would be even less if talks from 'external' speakers brought in as representatives of large high-profile experiments were excluded.

they are bound not to exist. It is indeed correct that unicorns are mythical beasts – but kangaroos are not. The only way to find out what is true is to go and look.

The final common objection is that it is too early. On this view, we should not start thinking about observational consequences of string theory until we understand what string theory is. According to this objection, before we try and connect string theory to observations we should first wait until we understand the structure of the theory better. It is premature to try and say what the theory predicts until we know what the theory is. We should instead continue to study the theory more, and more, and more, until we finally know exactly what it is. Then, and only then, should we start to think about empirical consequences.

The problem with this objection is that it is a time-invariant statement. It was made thirty years ago, it was made twenty years ago, it was made a decade ago and it is made now. It is also, by observation, an objection made by those who are uninterested in observation. Muscles that are never used waste away. It is like never commencing a journey because one is always waiting for better modes of transportation, and in the end produces a community of scientists where the language of measurement and experiment is one that may be read but cannot be spoken.

CHAPTER 11

Why Strings? Quantum Gravity

'Mummy, when I grow up I want to work out the theory of quantum gravity'. It is unlikely that any child, however precocious or driven, has ever uttered these words. However at a slightly older age this sentiment has been felt by many, many students of physics. Undergraduate physics is a wonderful time: you are hit by one deep insight after another, and over a period of several years you see the entire tapestry of the subject unrolled in front of you. The fusion of quantum mechanics and gravity is known to be missing from this tapestry, and in the full immortal confidence of youth it is easy to visualise yourself as the master weaver casually filling in the absent threads.

The statement of this problem was reviewed in chapter 4. We have a classical theory of gravity, general relativity, that was constructed by Einstein a century ago. This correctly describes gravity on large scales where we do not expect quantum mechanics to be important. While this is all well and good, we know that the world is in truth described by quantum mechanics. We do have working examples of quantum theories that describe the strong, weak and electromagnetic forces, and we expect that this should extend to gravity. What is needed is a quantum theory of the gravitational force, which will reduce to general relativity at large distances but at short distances will provide a truly quantum theory.

How short are these distances? The quantum nature of the electromagnetic, strong and weak forces is manifest by the time we reach distances around a billion times smaller than the size of an atom. However the estimate for the length at which quantum mechanics should become important for the gravitational scale is vastly smaller. This length is called the Planck length and is as small compared to these scales as an atom is to the great city of London. It is true that this estimate should be regarded as a worst case scenario, but there are also no known reasons for believing it badly wrong.

To guarantee being able to see quantum effects of the gravitational force, it is necessary to have technology that is able to probe such length scales.

No such technology exists. There are no known ways to make direct probes, not just of the Planck length but even of any lengths remotely comparable to it. The quantum mechanics of the gravitational force is well quarantined from the reach of direct experiments: even if you had the correct theory, how could you ever know that it was right?

It should be said that the quarantine of the Planck scale from experiment is not total. The absence of direct probes of quantum gravity does not preclude indirect probes, where physics present at the Planck length can bubble up to produce observable effects in doable experiments. While one very indirect probe was described in the previous chapter, some transmission is possible even for less indirect probes. The reason this can happen despite the smallness of the Planck length is the extraordinary accuracy to which many aspects of standard physics are known.

Examples are effects that violate special relativity, which is so well tested that even small violations of special relativity at the Planck length can be constrained and excluded. To give a more specific example, one way such effects could manifest themselves is through a dependence of the speed of light on the energy of photons, causing photons of different energies to travel at different speeds. This would be inconsistent with special relativity. If it were true, light from faraway sources in the universe would arrive at different times depending on how energetic the light was. This effect has not however been observed, thereby constraining the structure of physics at the smallest scales.

While there do then exist some probes of physics at the Planck scale, these probes are both limited and indirect. The field of experimental quantum gravity is, if not entirely barren, not fertile either, and it can only support a small and limited number of workers.

For all their early dreams, it is these hard facts that divert most of these keen young students away from quantum gravity. Comparison of theory and experiment is not the only aspect to science, but it is enormously fun. It is the opportunity to be either right or wrong, and the ability to know the difference, that gives science its vibrancy and its piquancy. If you want to have a scientific career, you would do well to enjoy it. For many, the paucity of experimental data would render a career thinking about quantum gravity unfulfilling.

That said, experimental data is not the only reason to be interested in a topic. The quantum mechanics of gravity contains many rich conceptual questions. What is a black hole made of, and how many different ways are there to make the same black hole? Does anything special happen as you approach a black hole? What happened to space and time at the beginning of the universe? Can space itself tear and then reform? Are there fundamental quantum limits on the amount of information space can contain? What happens when particles approach each other at distances less than the Planck length? These may all be theoretical questions, but they are well posed and should admit of answers.

Interest in these and related questions has been one of the major reasons to study string theory – and string theory has in turn responded through provocative answers that provoke more questions. This is the subject of this chapter, which focuses on three particular issues to do with quantum gravity.

11.1 COUNTING BLACK HOLES

We have already encountered black holes several times in this book. They are special objects that are relevant for both the most abstract parts of formal theory and the grubby world of gastrophysics. They occur when there is so much matter in so small a region that all the matter collapses in on itself. This collapse produces an object that is so dense that even light cannot escape from it: a black hole.

Black holes appeared in general relativity as a theoretical curiosity, when the German astronomer Karl Schwarzschild discovered the first example, the Schwarzschild solution, while serving in the German army during the first world war.[1] He died a year later, though, and relatively little attention was paid to this work for several decades. As interest revived, it would take until the 1970s before black holes began to be accepted as parts of the real universe: objects that actually existed in our galaxy and were in fact rather numerous. One of the first good candidates was the X-ray source Cygnus X-1, first discovered in 1964, over which Stephen Hawking had a bet with his friend Kip Thorne from Caltech. Hawking bet that Cygnus X-1 was not a black hole; Thorne that it was. The bet was four years subscription to the British satirical magazine *Private Eye* versus one years subscription to *Penthouse*. Hawking conceded in 1990, paying up 'to the outrage of Kip's liberated wife'.

Black holes in the galaxy are characterised by how heavy they are and how fast they are spinning. The typical mass of a galactic black hole is similar to or slightly larger than our sun. This is not surprising once you learn that they are usually formed from the inward collapse of large massive stars as they approach the end of their lives. There are much larger black holes at the centres of galaxies: our own Milky Way galaxy hosts a black hole at its centre with a mass of around one million suns. Such obese 'supermassive' black holes start small and then grow through consuming progressively more and more matter during the course of billions of years. Over this period, they turn into the galactic version of Michelin Man. There is typically one supermassive black hole per galaxy, playing a crucial role in galactic dynamics, as it is the centre around which all else revolves.

Beyond their role in the digestive processes of the galaxy, black holes are of interest in themselves. As discussed in chapter 8, the theoretical study of black holes reveals that they have a temperature. They are not entirely black. Not only do they absorb light, but they also emit it. Like any oven, they radiate

---

[1] Given the condition of first world war trenches, this provides an unusually apt example of the normally trite aphorism that we are all in the gutter, but some of us are looking at the stars.

energy. As we saw in chapter 8, this 'Hawking radiation' is a consequence of quantum mechanics: in a classical world, a black hole is indeed entirely black.

The temperature of a black hole is inversely proportional to its mass. A small black hole is hot. A large black hole is cool. The black holes that are present in the galaxy, with masses comparable to the sun, are very cold – with temperatures under one billionth of a degree. However, if we could compress a cubic kilometre of rock into a black hole, the result would be pleasantly toasty, with a temperature of a few hundred billion degrees and a power output of almost a gigawatt. As the lifetime of such a black hole would exceed the age of the universe, a collection of them in orbit around the earth would provide a permanent and clean solution to the energy problem.

Black holes also have an intrinsic spin. The spin is a rough measure of the internal rotation – how fast the black hole is, in some sense, rotating about itself. This rotation causes observers outside the black hole to find themselves dragged around the black hole instead of just falling straight in, and the larger the intrinsic spin, the larger this effect is.

In principle, black holes can also have an electric charge. This is a charge in the conventional sense: if a positively charged proton or a negatively charged electron falls into a black hole, the charge of that particle is added to that of the black hole. However this is insignificant for black holes in the galaxy, as the mutual attraction of positive and negative charges causes any non-zero charge to be rapidly neutralised.

Mass, spin and charge: these are the three quantities that define a black hole. From the values of these three quantities, it is possible to work out the entropy of the black hole. 'Entropy' is in effect a name for the number of possible internal arrangements of the object, or even more precisely the logarithm of the number of internal arrangements. It implicitly tells you how many ways there are to reorder the innards while the externals remain the same. The entropy of the gas in this room tells you how many ways there are to rearrange the molecules while leaving the pressure and temperature unaltered.

While we may not know exactly what is inside a black hole, the entropy tells us the total number of options. This expression for the entropy is a known formula, and was determined by Stephen Hawking to be exactly one quarter of the area of the black hole's event horizon.[2] The event horizon is the surface in the space around the black hole that marks the point of no return. Once you pass it, you are pulled unavoidably into the singularity at the centre of the black hole.

This poses a clean challenge to those who like clean problems. What are these internal arrangements (sometimes called *microstates*)? In a good theory of quantum gravity, clearly it should be possible to list these arrangements, count them, and confirm that they come in precisely the multiplicity

---

[2] As a point of technical pedantry, the dimensions of this statement are correct only in 'fundamental units' where Planck's constant $\hbar$, the speed of light c and the Planck mass M_P are all set to be equal to one.

anticipated by Hawking's formula. Equally clearly, this result does not tell us about any experimental observable: it would be career-shortening to try and look inside a black hole and see what it is made from.

However, it does offer a precise question with a known answer, and an ability to get the answer right is a necessary part of any candidate theory of quantum gravity. As well as a consistency check, a good understanding of the innards of a black hole would also offer considerable conceptual insight into quantum gravity.

The best place to ask these questions is not for the actual black holes that are formed within the galaxy. Even from a theoretical viewpoint, the black holes that actually exist within the galaxy are messy. As the problem of understanding black hole entropy is a theorist's problem, so it is best analysed in a theorist's playground: an idealised universe that is the mathematical version of a laboratory cleanroom. As occurs often in string theory, this idealised universe is one with large amounts of supersymmetry. Supersymmetry both simplifies the calculations and protects the results by ensuring that many otherwise dangerous effects cancel to zero.

While this question of understanding black hole entropy had existed in the background for a long time, for string theory the real breakthrough occurred in January 1996. In that month, Andrew Strominger of Santa Barbara and Cumrun Vafa of Harvard published a paper showing how they could reproduce Hawking's formula for the black hole entropy directly from a string theory calculation. The black holes they considered were certainly not 'realistic'. Not least, they existed in an imaginary world with four and not three spatial dimensions, where Strominger and Vafa had made only five of the ten dimensions of string theory very small. The black holes also belonged to a special class of black hole called extremal black holes. Extremal black holes are a type of black hole where the effects of its charge are, in a sense that can be made precise, exactly as large as the gravitational effects of its mass. As mentioned before, this certainly does not hold for the effectively chargeless black holes in our galaxy.

The property of extremality leads to several simplifications within the calculations. One reason Strominger and Vafa had to work in a world with four spatial dimensions is that, in a world with three spatial dimensions, extremality causes the area of the black hole horizon to go to zero. It vanishes entirely, and in principle you could approach as close as you dared to the black hole singularity without necessarily falling in. As the entropy is one quarter of this area, the entropy for such black holes vanishes – at least in three spatial dimensions.

The advantage of working in the fictional world of four spatial dimensions is that this statement is no longer true. In such a world, it is possible to have black holes that are both extremal and have a finite entropy. The advantage of extremality means that calculations are easier to do and under greater control. The advantage of finite entropy is that there is a meaningful target to try and reproduce – being able to reproduce an answer of zero is not hard and generally shows very little.

In this fictitious universe, Strominger and Vafa did two things within their paper. The first was to describe the five-dimensional extremal black holes they were using, work out the area of their event horizon and determine how it depended on the properties of the black hole. This part of the calculation was old: the techniques used descended from Hawking's original work. What this gave them was an expression for the entropy that they could then try to reproduce. The second part was the breakthrough, which was to explain microscopically from string theory where this entropy came from by counting configurations.

They did the counting using D-branes. We met D-branes in chapter 5: extended membrane-like objects of two, three or more spatial dimensions. Although always present in string theory, it had taken a long time for their import to be realised. It is also true that D-branes are heavy objects that are charged under analogues of the electric force. The large amounts of supersymmetry present in Strominger and Vafa's configuration made the D-branes extremal objects: appropriately measured, their charge and their mass are one and the same. As you put D-branes together, you assemble an object with both mass and charge. If you bind enough D-branes together, this object turns into a black hole.

How many ways are there of binding D-branes together? In the ideal universe of Strominger and Vafa, five of the extra dimensions are curled up. The details of the counting lie in the details of the curled-up geometry, and this curled-up geometry has many complicated properties. Asking how many ways there are to arrange the branes is a bit like asking how many ways there are of looping an elastic band around the quills of a porcupine: it requires both knowing the precise geometry and then counting carefully.

Strominger and Vafa had chosen their geometry carefully, and they were able to do the counting. The answer they found was precisely that of Hawking's formula – the number of configurations gave an entropy of exactly one quarter of the area of the black hole's event horizon, as required.

It is true that the black holes studied by Strominger and Vafa were artificial. It is not just that the dimensions of space were wrong. The calculations were also performed in a limit in which the interaction strength of strings was precisely zero. The world in which the calculations were done looked nothing like the real world of observations.

Nevertheless, for the first time there existed a calculation which explained, in a microscopic fashion, precisely where Hawking's entropy formula came from. It enumerated all possible innards of a black hole: and that enumeration added up exactly to what Hawking's formula required.

In the quest to understand quantum gravity, this result belongs to the good side. While clearly a triumph, it also leaves many questions to be explored. For example, what happens to black holes that are not extremal? Can the result also be extended to these black holes – and what about black holes in different backgrounds, black holes that are rotating, black holes that exist in smaller numbers of dimensions? There are many ways to take this result and extend it, and in doing so obtain an even better understanding of the quantum mechanics of black holes.

Since 1996, there have been something like one thousand papers that have done precisely this, using string theory to understand better the quantum mechanics of black holes from many directions. I am going to pick out one of these directions to focus on, which concerns subleading corrections to Hawking's formula for the entropy of a black hole.

One of the things that makes the entropy of black holes an attractive topic for a theorist is that it connects both the physics of the very big and the physics of the very small. Classical black holes are large objects – indeed, there is no real limit within general relativity on how large they can be. They are large enough that the space outside them is classical, with no large perturbations from quantum effects. One does not need to know quantum gravity to determine the entropy of a black hole. The classical, large-scale macroscopic physics is sufficient to compute the entropy, and this is precisely what Hawking did in 1974. However, we do need to know quantum gravity to write down the microscopic configurations that the entropy actually counts, and this is where Strominger and Vafa's contribution lay.

In fact, Hawking's computation of entropy can be improved slightly. In the great tradition of physics, Hawking computed not the full expression for the entropy of black holes but only the most important part. We recall that Hawking's formula was that the entropy of a black hole equalled one quarter of the area of its event horizon. The more massive the black hole, the larger its area, and the more true this result is. However, Hawking's result is not exact. There are additional contributions to the entropy. These grow not with the area but with the logarithm of the area. These are less important – for an event horizon area of one hundred, one thousand and one million in some arbitrary units, the logarithm would be respectively two, three and six – but they are still there, and they can still be calculated.

Furthermore, they can also be calculated using methods that do not rely on knowing the full details of quantum gravity. Just as for Hawking's original result, the subleading logarithmic corrections to a black hole's entropy can be determined knowing only the almost classical physics of big black holes, with no need to know quantum gravity. A better expression for the entropy of a black hole is then one quarter of the area of its event horizon – plus a small correction that depends on the logarithm of this area.

What about the microphysics? Strominger and Vafa had reproduced the first term in the entropy formula – entropy grows with one quarter of area. What about the subleading part that depends on the logarithm? Here the calculations are much harder, but for several cases these logarithmic terms have also been successfully matched. This gives an even more acute check of the fact that in string theory the microscopic counting of the number of states that can contribute to the black hole agrees exactly with the determination of the entropy through large-scale, semi-classical calculations.

This is further evidence that the formalism works and that string theory really can give a correct description of the innards of a black hole – even if so far only for black holes that exist in mathematical universes.

11.2   SINGULARITIES AND TOPOLOGY CHANGE

It is a truism that the arena for physics is spacetime. Events take place in space and are counted by time. Spacetime is the backdrop against which we record all that happens – but it is also a dynamical backdrop. In Einstein's theory of general relativity, the geometry of space and time themselves change under the influence of matter and energy. The more matter present, the more the geometry is deformed.

How far can this be taken? As the density of matter and energy increases, the geometry becomes more and more deformed – and eventually it breaks. In the language of general relativity, the geometry develops singularities. What this means is that within the equations that define the geometry, infinities start appearing. Sometimes in physics, infinities appear in equations that are not real infinities; they are simply an artefact of writing an equation in the wrong form, or choosing the wrong coordinates, and they disappear when the equations are rewritten correctly. The geometric singularities are not of this form. These infinities are real; spacetime becomes, according to the equations of general relativity, infinitely curved.

In this limit the equations of general relativity break down, signalling that Einstein's theory is no longer valid. This does not mean physics breaks down. Theories that fail outside their realm of validity are a common theme in physics. Newtonian mechanics is not a good description for objects that move at speeds close to that of light, and classical physics is not a good description of atoms. An important task for any quantum theory of gravity is to understand how it can resolve these singularities. In a correct theory, we expect to get finite, sensible answers in the region where classical general relativity gives nonsense. As geometry changes, we should be able to follow it to the places where general relativity fails – and then back out again.

There is a related question that one would hope quantum gravity would answer: can space tear and reform? In general relativity, geometry becomes dynamical – but this does not mean that anything goes. For example, it is impossible in general relativity to change the topology of space and add handles or holes to it. Topology remains unaltered under smooth changes – which are the only changes allowed by general relativity. The route to modifications of topology is barred by singularities, and it is only a better theory of gravity that can breach this barrier. These questions are not questions of observation or experiment; rather, they are classic questions of quantum gravity and good reasons to think about string theory.

String theory has had several clear and uncontrovertible successes in this area, replacing singularities present in general relativity with smooth behaviour. As the geometry becomes singular, nothing happens in string theory.

How and why is this so? The most prosaic answer is that string theory is governed by the equations of string theory and not by the equations of general relativity. If you examine these equations as geometry moves towards (certain) singularities, nothing happens. For certain classes of singularities, technically called orbifold singularities, these equations remain just as harmless at the singularity as they were for a smooth, featureless geometry.[3] This was recognised in the middle of the 1980s as work on string theory exploded.

There are other classes of singularities, called conifold singularities, which throughout the 1980s also appeared inconsistent within string theory. For these, even string theory did not appear to make sense. It was realised in the middle of the 1990s that this was due to a failure to include the effects of D-branes, whose importance had only been recognised in 1995. With the inclusion of D-branes, the conifold singularity also made sense in string theory.

While I have stated in bald prose that these two singularities are resolved by string theory, I have not explained why string theory should be different from general relativity: what comes in to save the day?

What is special about string theory? What is special is that it involves strings, and strings are extended objects. Some of the problems of singularities are associated with infinities concentrated at a point, and having a probe that is smeared out helps dilute the effects of these infinities.

However, there is a more precise and technical way that string theory helps de-smear the singularities. This is true for both orbifold and conifold singularities – although I warn that the following discussion may become a little difficult. A geometry approaching a singularity can be regarded as similar to a needle being progressively deformed so that it becomes ever sharper. The closer you get to the singularity, the sharper the point, and at the singular limit the point is infinitely sharp. The transition from a blunt needle to an infinitely sharp one is like the transition from a smooth geometry to a singular one.

The next step is to imagine a circular cross-section through the needle approaching the point. You see a circle that tapers and tapers in size, and it eventually reaches zero size only if the point is infinitely sharp. Now consider a tiny elastic band wrapped around the needle. The tension of the band contracts it around the needle. Far along the needle, the band still has a finite size coming from the width of the needle. For the singular geometry, at the point of the infinitely sharp needle, the contracted string shrinks to zero size.

---

[3]The simplest examples of orbifold singularities arise from taking a plane and identifying all points related by certain rotations – for example, identifying all points that are related by a one-hundred-and-twenty degree rotation. This leaves one fixed point, namely the coordinate origin, and it turns out that this fixed point of the symmetry has a singularity at it.

The string has zero energy – and so, according to Einstein's identification of mass and energy, zero mass.

The key point to draw from this analogy is that geometric singularities in string theory result in new massless particles as strings (or branes) are able to contract down to zero size. This occurs only because string theory is a theory of extended objects whose mass-energy is given by tension times length. The presence of these new massless particles in the singular geometry is what distinguishes string theory from general relativity. Although it is by no means obvious, it is precisely the existence of these new massless particles that turns out to cure the infinities that these singularities produce within classical general relativity.

String theory is then able to excise some of the geometric singularities from general relativity. While general relativity does not make sense on these spaces, string theory does. Where general relativity gave infinite answers, string theory produces finite numbers.

This is a good reason to work on string theory. It should however be said that string theory is not able to resolve every singularity. There are examples where general relativity gives singular behaviour – and it is unknown if, and how, string theory is able to smooth it out.

One example is the behaviour of singularities at the very beginning of the universe, so-called cosmological singularities. As we move the universe back in time, general relativity tells us that the universe becomes denser and denser and denser – and according to general relativity, at the very earliest time it was infinitely small and infinitely dense. This is a singularity that cries out for help from a theory of quantum gravity, and any such theory should certainly produce a sensible answer for this epoch. However plaintively it may cry out, string theory has so far not been able to help. Even in the simplest of toy models, there is so far no string theory account of cosmological singularities.

It is not that this is necessarily impossible, or beyond the ken of the theory – it is just that systems that evolve rapidly in time are harder to deal with than those that do not. No-one has ever been able to solve the equations or come up with an insightful shortcut.

However, the spatial singularities that can be resolved also lead to another beautiful result – the topology of space itself can change smoothly in string theory. As has already been mentioned, topology refers to the properties of spaces that remain the same under any smooth change. Topology – by definition – remains the same no matter how much a geometry is pulled, twisted or contorted.

Topology can be modified in two ways. One way is by tearing something apart and then reforming it, as with children using playdough who turn necklaces into figurines. This first way is not realised, as far as we know, in string theory.

The other way is by realising that singularities can sit at the border between topologically distinct spaces. Two different geometries, topologically distinct from each other, can have the same singular geometry sitting as a gate between them. Provided you can pass through this gate, the topology of space can be changed.

However, the singularities at the gate are precisely the ones string theory can deal with. Resolving these singularities allows the very topology of space to be changed: from a geometric perspective, it is possible to go into the singularity from one direction and with one topology, and come out again in another direction with another topology. This provides a controlled realisation of topology change, which is attractive as reasonable expectation would suggest that topology change should be possible within quantum gravity.

In summary then, string theory has had a lot of success in resolving singularities in space but less success in resolving singularities involving time. This provides inspiration from the past and open problems for the future.

11.3   THE ULTIMATE COLLIDER

Sitting outside Geneva, CERN's Large Hadron Collider is the most complicated machine ever built by humankind. It is a ring of superconducting magnets twenty-seven kilometres long, cooled to a temperature that is colder than space itself. Circulating protons within this ring, it can accelerate them to energies of six and a half trillion electronvolts in both clockwise and anticlockwise directions, before smashing them head on. The overall energy in the collision is thirteen trillion electronvolts. The volume around the collision vertices are surrounded by enormous multi-storey detectors, packed with wiring and sensors that record the details and debris of each event.

In terms of colliders that we can build, this is currently the best of the best. The growth in collider energies has also been impressive. Despite the technological demands, the energies involved in collisions at the LHC are over a thousand times greater than those that could be attained fifty years ago. Despite timescales of decades to design, excavate, build and operate colliders, their energies have continued to follow a particle physics version of Moore's Law, the computing rule of thumb that says that processing power doubles every eighteen months.

How long can this continue? The energy of current collisions at the LHC is smaller than the Planck energy by a factor of almost a million billion. Suppose, cometh the revolution, that the new world government decides to devote all resources to the construction of the PanTerran collider: an accelerating ring of magnets that stretches the entire way around the earth, looping from the North Pole to the South Pole and then back again. As ambition should be tempered with caution, we assume that the magnets used in the PanTerran collider are no stronger than those used in the LHC, and that as in the LHC the PanTerran collider is used to collide protons with protons.

The energies attained in the collisions would then be around ten quadrillion electron volts. This is a factor of one thousand larger than that occurring at the LHC, but still almost a factor of one trillion smaller than the quantum gravity scale. Even the use of alien technology to increase the size of the collider to the circumference of the earth's orbit around the sun would leave collision energies still a factor of a million smaller than the quantum gravity scale.

We see that collisions at the Planck scale are a long way off. This is not to say that it will never be possible to probe the quantum gravity scale directly. Only fools and fortune-tellers bet against what may be technologically possible centuries from now. However, it will not be in the school textbooks in the near future. What happens when two particles are collided with Planck scale energies is a definite fact of nature – but it is also a fact that is beyond observational ken anytime soon.

It still remains a good theoretical question. What does happen when elementary particles are collided with Planck scale energies? Even more specifically, what happens when gravitons are collided with Planck scale energies? Gravitons are the elementary quanta of the gravitational field, being to spacetime what photons are to the electromagnetic field. They are the basic, irreducible, elementary quantum excitations of that field.

As the minimal perturbation of gravity, gravitons are at the furthest remove in the subject of quantum gravity from ideas such as black holes or topology change. While black holes involve large objects that make large dents in spacetime, the graviton is the simplest possible excitation. Comparing a graviton to a black hole is like comparing an individual photon to the magnetic field produced in a tokamak. Both are configurations of the field, but one is large and classical while the other is small and quantum.

We have never measured a graviton – ever. Gravitons interact through the gravitational force, and so unlike photons the effects of individual gravitons are far too small to be detectable. This can be understood by thinking about how hard it already is to detect the gravitational pull even of a large bowling ball – and then extrapolate down to a single particle.

However, it is a perfectly valid theoretical question to ask what would happen if two gravitons approached each other with energies close to that of the Planck scale. How would they interact and what would the possible results be? This question is well defined. It involves the quanta of the gravitational force. It also involves something that – in principle, if not in practice – could occur in this world.

This question is also interesting because general relativity does not give a good answer to it. For collision energies much less than the Planck scale, general relativity does allow us to compute what will happen when two gravitons approach each other. However, as the energies increase general relativity ceases to be a predictive theory. This is because general relativity is, in the parlance of chapter 3, a non-renormalisable theory.

As described there, for the quantum field theories of the Standard Model, the techniques of renormalisation allow for a finite number of infinities to be isolated and eliminated. For such renormalisable theories, it is possible to trade the appearance of infinities in calculations for measurements; by measuring enough quantities we can remove the infinities. The quantum mechanics of general relativity leads to a proliferation of infinities – indeed, it leads to an infinity of infinities – and these cannot be eliminated via renormalisation.

For gravitons that approach each other with Planck scale energies, general relativity is not a predictive theory. It is impossible to use general relativity to work out what will happen – for what is clearly a well-defined physical process. A clear task for quantum gravity is to be able to give finite, predictive answers for this process. Even if the experiment can never actually be performed, a quantum theory of gravity is required to provide an answer.

Historically, this was one of the leading motivations for working on string theory. The fact that string theory does give finite answers to this question has actually been known since before string theory was recognised as a theory of strings. In this case, the answer was known even before the question was posed. As we saw in chapter 5, in string theory graviton particles exist as oscillatory modes of a closed string. The question of how one can compute the scattering of gravitons is the same as the question of how one can compute the scattering of strings, and formulae for the scattering of strings existed even before 'string theory' was recognised to be a theory of strings.

It is true that in string theory, the scattering of gravitons with energies close to the Planck scale is finite. By itself, this is only so interesting. However in string theory there is another distinctive feature of the scattering of gravitons at these energies: this is the *softness* of the scattering. As described in chapter 5, the scattering of objects can be viewed as either hard or soft. While there is not a rigid definition, hard scattering is the type of scattering produced by ball bearings. Ball bearings colliding at high velocity can easily ricochet and fly out again at large angles to their original direction of approach. Soft scattering is that produced by blobs of jelly, which are incapable of ricocheting and either disintegrate or continue broadly in their initial direction of motion.

When strings are collided, the equations of string theory tell us that they are more like blobs of jelly than ball bearings. Two strings, heading for each other at sufficiently high energies, are extremely unlikely to come off at right angles, and the more energetic they are, the less likely this is to happen. Instead, they can only be deflected by small amounts. Their scattering is soft, and it can only divert the strings by minor perturbations.

This is a characteristic feature, and a characteristic prediction, of string theory: when our far-off descendants build the Ultimate Collider and use it to smash together particles with energies at the scale of quantum gravity, they will find that these particles interact as strings and are therefore unable to scatter by large amounts.

This is clearly a prediction – but a theoretical rather than a practical one. Those who work on the applications of string theory to quantum gravity do not generally do so because they are hunting for falsifiable predictions and seeking an imminent confrontation with experimental data. Instead, they care about quantum gravity and the types of questions that arise when one tries to combine quantum mechanics and gravity. These questions are well removed from direct empirical probes, but no less valid for that.

This chapter has discussed three aspects of quantum gravity, and over the last four chapters we have now seen several good motivations for working on string theory. There are many different styles of doing science successfully, and this variety of topics appeals to the different varieties of physicist. We now turn from these technical topics to a more light-hearted discussion of the different kinds of physicists and the many ways of doing physics.

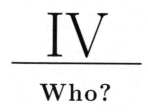

IV

Who?

CHAPTER 12

A Thousand Flowers Blooming: Styles of Science

Over the four previous chapters, I have described some of the many different reasons for caring about string theory – what do all those people who work on 'string theory' actually do? The diversity of topics reflects a diversity of interests. The diversity of interests reflects the diversity of individuals. It also reflects very different styles of doing science.

These different styles are the topic of this chapter. The previous chapters have been heavy, and this is intended to be lighter. It will provide idealised caricatures or portraits of common examples of the genus theoretical physicist. None of these caricatures represent actual individuals. Everyone is unique, but this uniqueness co-exists with certain styles of approach to the subject that are common to many.

There is also a more serious purpose to these descriptions. Today the route to professional science starts out with the PhD degree. Eighty years ago, it may have been possible to have a scientific career without a doctorate, but those times are gone. However, the PhD is about more than just solving problems. For the aspirant student, an important part of this process is to discover what type of scientist they are and what makes them happy. Different styles are suited to different problems, and the subject advances most rapidly when square pegs fit into square holes. Like an author finding their mature voice, for young scientists independence comes from finding the niche and style that suits them.

The descriptions I offer below deliberately contain elements of carica-ture and exaggeration. They are also not always maximally respectful. There are several good reasons for this. The first is that as a rule satire is more truthful than hagiography. The prints of Hogarth are more instructive about

18th century London than the marmoreal memorials lining the walls of Georgian churches.

The second reason is that the subject contains enough trumpet blowing as it is. It is genuinely true that quantum field theory and quantum gravity are hard and important subjects. It is genuinely true that in order simply to understand these topics, and even more so to contribute to them at a meaningful level, one needs a level of analytical intelligence that is higher than average. It is also true that a prerequisite for a career in the subject is a solid work ethic and a healthy dose of self-confidence – indeed, most young physicists start with the secret hope of contributing at the level of Einstein, Dirac or Feynman.

It does not follow that everyone who has proposed an idea that is not immediately ruled out is therefore a genius. Both an overly large self-regard and a feeling of being set apart from one's fellow citizens feature heavily in the *deformation professionelle* of theoretical physics. It is too easy to think that, through contemplating the laws of nature, one belongs to a higher, more exalted plane of existence than those whose sweat and taxes fund this research. The conviction of genius is commoner than the reality, and physicists solving problems have more in common with sewage engineers clearing blockages than they care to admit.

The third – and best – reason is that physics is done by physicists, and physicists are human. We all have different paths through life, and these are reflected in the varied approaches to physics. Physics is taught largely as an impersonal tale of results in a textbook. In the real world of active research on open problems, personality enters both in the problems people choose to address and the way they go about attacking them. A range of questions benefits from a range of physicists, and without a range of approaches the subject would stagnate. Einstein was a different kind of physicist than Dirac, who was a different kind of physicist than Feynman. No one type is 'better' absolutely, anymore than a spanner is 'better' than a hammer which is 'better' than a screwdriver.

Anyone who holds that science is a social construct is nuts. However anyone denying that social factors do in part determine which topics are regarded as important is also nuts. In the long term, experiment is the great and impartial judge that depersonalises the subject. In the short term, if you want an honest understanding of the state of current research, then it is good to have a sense of the types of people who do it.

12.1 THE REVOLUTIONARIES

There are few subjects as mythological as the history of science as taught by scientists. The conventional telling of scientific history is a wonderful story. It is a story that crosses hundreds of years and involves diverse languages and cultures. It is a story of brave and unconventional thinkers who have overturned the dead hand of authority through their own independence of mind.

It is a story of heroes. There is Christopher Columbus, who set off westwards to reach the East Indies in defiance of the flat-earth dogmatists of the Spanish court. There is Galileo, surrounded by sour-faced cardinals ordering him to recant, but still softly murmuring '*Eppur si muove*'. There is Isaac Newton, struck by a falling apple and being inspired to his insight of universal gravitation. There is Charles Darwin and his bulldog Thomas Henry Huxley, putting forward and defending evolution in the teeth of opposition from a clerical establishment insisting on the inerrancy of scripture. There is the shock white hair of Einstein and the joker Richard Feynman – what do you care what other people think? On this account, the main characteristic of a good scientist is conceptual originality and contempt for authority. This history is magnificent. This history is inspiring. This history is also, sorry to say, bunk.

It is true that almost all discoveries of the first rate fly in the face of conventional wisdom and are contrary to the generally accepted theories of the time. This truth is, however, tautological. A result that only confirms that which everyone already suspected to be true is, entirely reasonably, less important than one which refutes it. The discovery of the Higgs boson at the Large Hadron Collider was widely anticipated. It was expected to happen. It did happen. The result was important – good champagne was uncorked, and Peter Higgs and François Englert got the Nobel Prize. However, it was not revolutionary. The Higgs boson was the last missing ingredient of the Standard Model of particle physics, and its absence would actually have been more surprising than its presence.

The converse is not true. Personally, I am so far a moderately successful physicist with a decent career but no earth-shattering results or any broader name recognition. However, even I still receive several times per month accounts by email of amazing theories that correct the errors of Newton, Einstein and Maxwell, and on occasion all three simultaneously. These theories promise high and deliver low, and they are wonderfully and bizarrely bonkers.[1] Regrettably, the fact that an idea is believed nonsense by the scientific establishment and contradicts hundreds of years of theoretical and experimental spadework is not sure and certain evidence that the author can book his tickets to Stockholm.[2]

Setting aside crackpots, it is still true that a lot of scientific work is drudgery. Drudgery is perhaps an unfair word. The majority of work in theoretical physics requires careful and accurate calculation over a period of months, years or decades. The Standard Model is an example of a quantum

---

[1] The most truly batshit crazy of these 'theories' double down on their ideas by merging revolutionary accounts of physics with political conspiracy theories. The results have to be read to be believed; I will forward examples of these theories to anyone interested.

[2] It always is a he. While most physicists are male, there are many outstanding women in the subject. Crackpots are however universally men – I have yet to encounter even a single female crackpot.

field theory, and as discussed in chapters 3 and 8, quantum field theory is complex and has many subtleties.

The great value of physics, and what sets it apart from many other disciplines, is the ability to predict. But, when working out these predictions it is not easy either to obtain the correct answer or to know the answer is correct once it has been found. The ability to predict is a byproduct of the ability to calculate, and the ability to calculate is a byproduct of the ability to apply, for an extended period, the seat of the pants to the seat of the chair.

Most of the time such calculations only extend the known validity of pre-existing theories such as the Standard Model. This work is important, but in a sense routine. The basic rules are known, and the job is to apply them correctly. It entails accepting the job of cook rather than captain on the Mayflower. You cannot fail spectacularly, but the chances for glory are also limited.

Calculation is also a reminder of the collectivist nature of much of physics. The subject is built up through the labours of many people over many years. The underlying equations of the Standard Model of particle physics were fully written down by the middle of the 1970s. To a divine and omniscient mind, all its logical consequences would have been immediately visible.[3] To us, they are not. Obtaining any sort of sight of them requires first many hours of labour and then again many more hours removing the errors that have inevitably crept in.

Some people rebel against this. Undergraduate physics is an unparalleled intellectual experience: it is a smorgasbord of the deepest and most powerful thoughts that have ever been thunk. You learn physics at a rate of a Nobel Prize a week, and the resulting frisson of the mind is at a level that is never experienced again. Four years takes you from Newtonian gravity through the laws of heat and energy, past Maxwell's synthesis of electromagnetism and into the laws of special and general relativity, from the basics of quantum mechanics to the Standard Model and quantum field theory. It is wonderful and magnificent. After all this, it is easy to feel a sense of *ennui* at the prospect of becoming merely a cog in the calculating machine. The siren call of Big Ideas is more attractive than the factory hooter summoning workers to the assembly line of computation.

This rebellion carries a dream, the dream of Galileo: to give the finger to the establishment with a new, revolutionary idea that makes previous conceptions redundant. A revolutionary idea needs an appropriate canvas, and this canvas can only be provided by the big questions: Is quantum mechanics correct? What is the correct theory of quantum gravity and what are the principles that govern it? What is the nature of the dark energy that appears to dominate the energy budget of the universe? Are there extra spatial dimensions with

---

[3]This fact has led to one of the worst possible pieces of advice provided for giving seminars – assume the audience has zero knowledge but infinite intelligence.

sizes possibly as large as a millimetre? Is there a multiverse, or landscape, of universes? Is the anthropic principle responsible for the 'constants' of nature?

These big questions offer the chance of altering the entire course of physics. They modify not just the details but, if answered correctly, the whole way one should think about the subject. These are areas where an idea that lasts will survive for decades and centuries, and enter the enduring legacy of physics.

In contrast, no such grand historical narrative is offered to ideas that aim at calculating next-to-next-to-leading order corrections to the production rate of a W boson together with three quarks in proton-proton collisions at the Large Hadron Collider. Big revolutionary thinkers need big revolutionary questions, and they recoil from such detail.

Science always needs a supply of would-be-revolutionaries. Most of the time, this approach only allows a researcher to sit on the periphery of the subject while looking down on those who toil in the vineyard. Just occasionally, however, there are discoveries to be made that can only be made in this manner. In all parts of life, true revolutionaries have despised the *petite bourgeoisie* who take the road more travelled by. In the case of real genius, the road less travelled by leads to glorious discovery. In most cases, however, it peters out to an empty trail in the middle of nowhere.

12.2   VORSPRUNG DURCH TECHNIK

'The thing that hath been, it is that which shall be; and that which is done is that which shall be done: and there is no new thing under the sun. Is there any thing whereof it may be said, See, this is new? It hath been already of old time, which was before us.' [Ecclesiastes 1:9-10]

It is not only in physics that an absence of new phenomena depresses the wise. As we saw originally in the first chapter, the Standard Model of particle physics is contemporaneous with the Vietnam War: it was developed during the 1960s and 1970s, and it was essentially finished by 1975. The theory of electroweak interactions was developed during the 1960s and shown to work as a quantum theory in 1971. This electroweak theory extended the account of quantum electrodynamics that had been developed in the late 1940s, merging it with a theory of the weak force. In a period of ten years from the 1960s to the 1970s, the strong force also went from being baffling to being understood. The final step was the discovery by David Gross, David Politzer and Frank Wilczek in 1973, encountered in chapter 1, that at shorter and shorter distances the strong force becomes progressively weaker.

The Standard Model was written down by bolting together these various different components. At the time, it predicted the existence and interactions of many particles that were yet to be discovered. When first put down on paper, there seemed no reason for the Standard Model to survive five years, let alone thirty five.

However, since then all the particles whose existences were predicted by the Standard Model have actually gone on to be found. The gluon, the force carrier of the strong interactions, was identified in 1977 in four separate experiments at DESY – Deutsches Electronen SYnchrotronen, the German accelerator complex on the outskirts of Hamburg. The tau lepton was discovered in 1975 at the Stanford Linear Accelerator and the bottom quark in 1977 at Fermilab. In 1983 the enormously heavy force carriers of the weak interaction – the W and Z bosons – were found by the Underground Area 1 collaboration at CERN, led by the elemental and volcanic force of nature that was Carlo Rubbia.[4] The W and Z, individually, are each approximately as massive as an atom of bromine, and at the time of discovery were twenty times heavier than any other known particle.

Since then, two still heavier particles have been discovered. The first was the top quark, around twice as heavy again as the W and Z particles, which was discovered in 1995 at the Fermilab complex near Chicago. The final ingredient of the Standard Model was the Higgs boson, which was discovered in 2012 at the Large Hadron Collider at CERN.

All this time, all these colliders – and the Standard Model has yet to crack. Envisioned as a cardboard shack, the Standard Model has proved Kevlar-plated.

Over this period, the quality of theoretical predictions has improved dramatically. The original predictions of the Standard Model were evaluated at 'tree-level' or 'leading order in perturbation theory': essentially just the first term in an approximation scheme. 'Leading order' is the spherical cow level of approximation – if you had to estimate the weight of a cow, the simplest method is to model the cow as a sphere. These rough approximations were then extended to 'next-to-leading-order in perturbation theory' – this is the approximation of an ellipsoidal cow. The most recent extension is to 'next-to-next-to-leading-order', at which point the cow starts sprouting legs and a face. The time period to move between each successive level of approximation is often around a decade for each step. If this sounds a little lazy, it is because the complexity of the computation grows enormously with each 'next-to'. If a tree-level computation has ten components, a next-to-leading-order computation may have five hundred parts and a next-to-next-to-leading order computation twenty thousand elements.

Over this period, the increase in the energies of particle colliders is equivalent to a decrease in the distances probed by a factor of two hundred. The statement that the Standard Model has continued to describe nature successfully across this range of distances is important. It tells us that the Standard Model works, and by doing so it significantly extends our understanding of nature.

---

[4]The year 1983 was a triumphal year for Rubbia's collaboration: they discovered not only the W^+, the W^-, the Z^0 and the top quark, but also supersymmetry. The following year was however less good as the last two had to be undiscovered.

It is also important that this statement is true, and so actually can be made. This statement can only be made because we know what the Standard Model predicts. It is only by knowing what the Standard Model predicts that we can compare its predictions to experiment. This knowledge has required work; it has taken tens of thousands of years of effort by thousands of individuals to determine and refine the predictions of the Standard Model to a level that can be tested in the progressively more demanding environments that newer and newer colliders have provided.

There are many aspects to this. There is the level of calculational technique – once a calculation involves many thousands of individual elements, it is necessary that the methods used are computationally economical and efficient. As seen in chapter 3, the techniques of renormalisation make the individual elements of the calculation formally infinite, with these formal infinities cancelling between different parts of the computation. It is important that the calculation is structured so that these infinities do manifestly cancel: a single stray infinity makes mockery of any claims to precision.

There is the level of verification: once a calculation is done, how do you know that the answer is correct? What tests can you do to ensure that you have not misled yourself, and that the answer is indeed the correct one? What are the equivalents of checksums – consistency checks that ensure the answers obtained are not simply meaningless?

There is the level of coding. Once calculations have gone beyond a certain level of complexity, they must be performed numerically and on a computer. Large codes need to be written to perform these calculations, and large codes also need to be maintained and debugged. Uninitialised variables and the conflation of signed and unsigned integers can render results just as meaningless as if the wrong laws of quantum mechanics had been used.

Even once a calculation has been done, it still has to be compared with experiment. Particle physics experiments can be many storeys high and they are packed with cables. Their outputs consist not of 'Higgs bosons', but of electronic signals. A primary energetic particle at a collider may manifest itself as a complex jet of hundreds or thousands of secondary particles, approximately collimated along the direction of the original particle. These hundreds of particles in turn deposit their energies in large calorimeters, producing electronic signals that record the locations and amounts of energy deposited. Further modelling is needed to turn the basic results of the calculation into the observables that a collider experiment can actually measure – and large codes are also needed to describe this.

Finally, there is even the level of data storage. The Large Hadron Collider can store only a minute fraction of all the collisions that occur. Most of the data never even makes it to memory, and elaborate 'triggers' are used to determine the collisions that get recorded and the collisions that do not. In comparing any prediction to data, one must account for the selection effect that over 99.999% per cent of the 'data' was immediately discarded.[5]

Despite all this, the predictions of the Standard Model are ultimately precise and well-defined. At the end, there is a theoretical number that can be compared to an experimental result. Determining these predictions requires large amounts of work at each of the many levels described above. For this to happen, it needs people – many people! – to devote their entire careers to working on single parts of this process. The physicists who follow these paths spend years calculating the precise implications of the Standard Model for regimes where it has yet to be tested. It involves turning the equations of the Standard Model into real predictions for real data.

Their work will never win the Nobel Prize. It involves working within a well-defined framework, in which there is a well-defined answer, and evaluating this answer. The rules of reward are simple. Nobel Prizes and the highest scientific honours are not given to those who refine the theories of others. Their results may be important. The calculations may be essential for confirming any observation of new physics. *Ce n'importe quoi* – the highest awards are reserved for those who develop their own frameworks rather than those of others.

Of course those who choose to work on, for example, the theory of the precise predictions of the Standard Model know the choice they are making. It is a choice not to work on the latest hot topic. The fashionable areas vary hugely with time. At one time it was proton decay and grand unified theories, at another time superstrings and Calabi-Yau compactification, at another time millimetre-sized extra dimensions, and for one very brief and very embarrassing instant it was neutrinos that could travel faster than the speed of light. The precise form of the higher-order corrections to the predictions of the Standard Model are never fully in fashion – but they are never out of fashion either. Some ideas – the existence of millimetre-sized extra dimensions is a good example – can go rapidly from the height of fashion to almost total death. The Standard Model will never suffer this fate and will always be of value.

This style of work is essential for progress in physics. The aim of these calculations is in general not directly the discovery of new physics – no one computation is done with the expectation that it will lead to a breakdown in the Standard Model. However, any such result may end up as the key element in a discovery. A minute, but established, deviation from expectation can be the herald of great change. For example, it was the anomalous advance in the perihelion of the planet Mercury, at a rate of one-hundredth of one degree per century, that was the first harbinger of general relativity.

---

[5]The details of the trigger are one of the terrors of the night for CERN experimentalists: what if there was a major fundamental discovery accessible to the Large Hadron Collider, but it was missed because the trigger caused all the necessary data to be placed in an electronic garbage can?

Something similar may in time be true for the Standard Model. For example, one of the four main experiments at the LHC is called LHC-B. It is designed to study the production and decay of particles involving the bottom quark. One way the bottom quark can join with other particles is to produce what are called B-mesons, consisting of a bottom quark and an antiquark. These B-mesons live for a short period of time and then decay. Among other goals, the LHC-B experiment aims to measure precisely the rate at which these B-mesons decay and the different types of particles they decay to.

Of particular interest are so-called 'rare decays' – processes that the Standard Model predicts will happen maybe only one time in a billion. If such a B-meson were found to decay in this manner even as much as once in a hundred million times, it would be an unambiguous signal that the Standard Model is wrong – a signal as unambiguous as if a new heavy particle were to be discovered directly and spectacularly. If this occurs – and while it has not happened yet, it could – it would only be because of long and careful calculations of exactly what the Standard Model predicted.

You will however find such calculations only rarely featured in the pages of *New Scientist* or *Scientific American*. Let it not be said that this has gone entirely unnoticed, or that those who work on the predictions of known theories ever feel a tinge of satisfaction at an absence of signals for new, speculative and well-hyped ideas. Let it not be said that schadenfreude is ever felt when such ideas fail, or that the trendy set's dismissal of precise calculations as 'German physics' could ever contribute to this emotion. Let it not be said, because of course it is not true.

I have chosen to illustrate this style of physics using the example of precision calculations within the Standard Model. However, it is not confined to this topic. Many areas – certainly including string theory – have definite frameworks, in which well-defined but difficult calculations can be carried out, leading to well-defined results. Examples in string theory are tests of the AdS/CFT or gauge/gravity correspondence, which I touched on in chapter 8. These tests involve difficult but precise computations in quantum field theory that are directly analogous to the next-to-leading-order or next-to-next-to-leading-order computations in the Standard Model. Over the almost two decades since the AdS/CFT correspondence was first formulated, these tests have become progressively more intricate and complex.

It is a very human choice to follow paths of clear scientific value even if the act of doing so may exclude the absolute highest levels of achievements. It may perhaps be compared to choosing to be an accountant rather than a dotcom entrepreneur. Both business and science need those who will risk everything, and those who prefer definite but limited return over a small chance of glory.

12.3   STOCKHOLM OR BUST

As we have also seen, the Standard Model is, in matters empirical, conceptual and practical, the very model of a modern scientific theory. It works. It is deep. It correctly predicts the results of a dazzling range of experiments to an enormous level of precision. However, every particle physicist also knows that the Standard Model is incomplete. It is not a full theory of nature. Every particle physicist knows that, at some energy scale, new particles and new interactions will appear. They know that there is a new theory, which lies beyond the Standard Model but encompasses it, and which is true in the deepest and most wonderful sense of that word. That theory is out there, waiting to be found, in the same way that still untouched hoards of Roman and Saxon gold lie in country fields waiting for the farmer's plough. They also know that glory is reserved for the first person to write that theory down.

This section is about model builders, those who propose ideas for new physics beyond the Standard Model that can then be tested experimentally. This sober description does little justice to the reality. There is a traditional story, popular in textbooks, of how progress in physics occurs. Those outside science like to call this story the scientific method, and it goes like this. Alice is a theorist. She thinks hard and invents a new theory about nature. Alice is aware of the results of existing experiments, and she checks that the new theory is consistent with the existing data. Alice then works out the predictions of the theory for measurements yet to be performed. Bob is an experimentalist. Bob looks at the predictions of Alice's theory. He thinks about what these predictions mean and builds an experiment to test them. He carries out the experiment and analyses the results. Do the results agree with the predictions that Alice made? If yes, bully for Alice. If not, her theory is wrong, another false idea is discarded and the great engine of Science steams on.

This is the theory of scientific progress, and in theory it should coincide with the practice. In practice, it does not. To succeed, this approach requires cheap experiments and a surplus of data compared to theories. However in the world we live in, ideas are quick and cheap but experiments are long and expensive. This can be seen by looking at the electronic paper repository arXiv.org. The arXiv – as it is known – has revolutionised scientific publication since its inception in 1992. Every weekday, fifteen or so papers are uploaded to the phenomenology section of the arXiv preprint server, the section concerned with theories that may explain data. If we take twenty pages as the typical length of a paper, this corresponds to the uploading of one medium-size book a day. Discounting weekends, fifteen papers a day adds up to three hundred papers per month. Even if simply one paper in every ten of these involved proposals for models of New Physics – and regular readers know the true fraction is higher – that still makes one such proposal a day, a rate far in excess of the build time of experiments.

Any scientist who proposes the existence of a new particle or a new force of nature is one dreaming of a big discovery. They are hoping, with a fortunate roll of the dice, to be in the right place at the right time and with the right idea. Unlike the would-be revolutionaries, they are not seeking to overturn the entire structure of physics. It is sufficient for them to be right once. It is somewhat analogous to the footballing phenomenon of the goalhanger, the player who always waits around the opposition's goal, hoping for the chance to stick the ball in the net.

The fact that the stated aim of model building is to make a prediction of new physics, and to see it confirmed by experiment, has an important corollary. Particle physics experiments take a long time to build, and the backdrop for this book is the fact that over the last forty years the Standard Model has been enormously successful at describing nature. The sad but sober truth is that over this period no evidence for any physics additional to the Standard Model has emerged. This statement can be glossed in various ways, and various small exceptions urged – what about neutrino masses? What about dark energy? What about inflation? While these points can be debated, it is not up for debate that the large accelerator complexes have provided fantastic confirmations of the Standard Model but no signs yet of any new physics beyond it.

It is striking that this history has in no way prevented many physicists from building careers – indeed, from building highly successful careers – through putting forward ideas for new physics beyond the Standard Model. Not a single one of these new model-building ideas put forward since 1974 has ever succeeded. This does not prevent the originators of these ideas receiving some of the highest accolades in science, as well as the largest financial rewards available.[6]

In terms of actually predicting novel experimental phenomena, then, none of these models have been successful. What does make a model successful? From a worldly perspective, a model is successful if lots of attention is paid to it. When this happens, postdoctoral researchers will add epicycles to it, while experimenters will make dedicated searches to look for its features and place limits on its parameters. Graduate students will develop further modifications, and write PhD theses explaining how the aforementioned experimental limits can be evaded by removing the twelfth epicycle and adding a sixteenth one. Large numbers of papers will cite the original incarnation of the model, thereby reconfirming the importance and significance of the original paper. As the absence of any positive evidence accumulates, models sublimate into infinitely flexible paradigms that can never be excluded and can accommodate any observation.

The scale of this can be impressive – although to be fair, one should not blame the original authors for the subsequent take-up of an idea. As an illustration, the four original papers of two popular models – the so-called Large Extra Dimensions and Randall-Sundrum scenarios – have around 5500, 3800, 6800 and 5500 citations respectively. These are enormous numbers for

---

[6]While the Nobel Prize committee insists on experimental confirmation of ideas, Silicon Valley billionaires do not. The largest monetary prize in physics is the three-million-dollar Fundamental Physics Prize founded by the technology investor Yuri Milner. One of the inaugural awards was for 'original approaches to outstanding problems in particle physics', even though the predictions of these original approaches have failed all experimental tests.

phenomenological ideas with zero experimental support.[7] Such models can rarely if ever be totally ruled out, as there is always some wiggle room for playing with parameters and tweaking terms so as to avoid any given experimental constraint. As models evolve into industries, pointing out the absence of any favourable experimental evidence becomes unseemly: a bit like commenting on the Emperor's willy rather than on the magnificence of his clothing.

To become a successful model builder, it is then not a requirement that the models actually describe nature. The overabundance of models compared to data has an additional effect on the language used in papers. As models need attention like humans need oxygen, the consequence is adjectival inflation. Models are natural, explanations are elegant, and conclusions are compelling. For the prospective model builder, salesmanship is a not unuseful skill to have.

In principle, all of the above makes zero difference. Science is driven by experiment. Over time, truth will out. Inflated claims are exposed, and nature, not us, will determine the correct explanation. Indeed, the Standard Model itself was in part built according to the above method. Claims were made, ideas were proposed – and data eventually singled out the ideas that were correct and dissolved inappropriate hyperbole. In the end, the models that are correct survive and flourish, while those that are wrong are winnowed away to the aether as chaff.

This is indeed all true – in the long run. In the long run, we are also dead. While the long run may be characterised by eternal verities, the short run is what determines the quotidian mundanities of jobs and salaries that can afford a mortgage. Flattery may not soothe the dull cold ear of death, but it is pleasing enough to the living.

12.4 THE MOST SUBLIME BRAHMINATE OF PRINCETON

Theoretical physics is a deep subject that concerns itself with big questions. As we have seen in this book, some of these questions are very big indeed and lie at the heart of quantum gravity. What is the fundamental physics of black holes? What is the nature of space and time? What is physics like at energies near the Planck scale? What is M-theory? These questions are hard, with no answers that can fit into a one-hundred-and-forty-character tweet. As we have also seen throughout this book, there will be no quick experimental solution. The ugly detritus of proton collisions at the Large Hadron Collider will not tell us about physics at distance scales smaller by a factor of ten to

---

[7]Funding councils that focus on metrics to evaluate quality may find amusing one aspect of the citation records of the two Randall-Sundrum papers. The arXiv numbering of these papers are hep-ph/9905221 and hep-th/9906064, referring to '(h)igh (e)nergy (p)hysics', and then either (th)eory or (ph)enomenology, followed by year, month and paper number. However, with referencing people sometimes get confused. There is a paper by different authors on a relatively minor and obscure topic, but with the numbering hep-th/9905221. This paper now has almost two hundred citations – but examining these it is clear they almost entirely come from people mistyping 't' for 'p' when attempting to reference the first Randall-Sundrum paper.

the fifteen than those that can be probed at CERN. How can these questions then be answered?

The history of western philosophy is glibly observed to be a history of commentaries on Plato and Aristotle. Plato and his student Aristotle were two of the greatest philosophers of the ancient world, but their approaches to philosophy are marked by two distinct styles. Roughly, Plato saw objects in the world as approximations to ideal objects, called Forms. When I use a pen to draw a circle on a sheet of paper, it is not a perfect circle. My hand shakes slightly and is not perfectly steady as it draws. The line has a finite width. Nonetheless, you can recognise that what I am trying to draw is a circle, and we can conceive of the notion of a perfect circle even though no such perfect circle has ever been drawn. According to Plato, the perfect circle exists as a Form, and the actual circle I draw participates in this Form. These Forms – the idealisations of actual objects – play a central part in Plato's philosophy. We are instructed to look beyond the imperfect reality of the pencil drawing of a circle on a page, to the Form of the perfect circle, of which we can conceive despite never having seen an explicit example. The focus of study is not the actual circle on the actual page, but the Form to which the drawing alludes.

With Aristotle, the focus is reversed. Aristotle was not a 'scientist' – the word was not even coined until 1834. He was however inordinately curious about the details of the natural world and interested in the haecceity – the thisness – of objects. Aristotle wrote about physics and the mechanical motion of objects. What makes a body move? Why does it continue to move and then stop? He wrote about biology – what are the functions of plants and animals? What does the oesophagus do? These were not just abstract queries, but questions based on lengthy personal observation of animal and plant life in its many possible forms. He also wrote about astronomy, poetry, ethics and politics – while additionally functioning as personal tutor to the future Alexander the Great. One cannot help reflecting that at a dinner party he would have been either the most sparkling company or a consummate bore.

The difference in style between Aristotle and Plato is roughly that while Plato saw an object as an imperfect version of a perfect Form, Aristotle saw the object in all its particular details: the hair of the cat, the teeth of the dog, the leaves of the tree. Seeing a snub nose, he did not extract an ideal property of 'snubness'; the property does not exist without the nose. To Aristotle, the details were not mere accidental features of an object – in a certain sense, they were the object.

The difference between the Platonic and Aristotelian approaches is the difference between those who want to remove inessential details to focus on the core properties of something, and those who see the details as the most revealing part. It is the difference between seeing the particular as an example of the general, and seeing the general as a collection of particulars. It is the difference in emphasis between loving humanity and loving your spouse. In particle physics, it is the difference between studying maximally symmetric super Yang-Mills theory and studying the actual strong force.

The world needs both Platonists and Aristotelians, and so does theoretical physics. In physics the *spiritus movens* of the Platonists is located a short walk from the dormitory town of Princeton, New Jersey, at the Institute for Advanced Study. The Institute for Advanced Study was founded in 1930 as a pure research institute. It is located in a calm location, surrounded by woods and with deer running onto the Institute lawns. With an endowment of over five hundred million dollars, it is an iconic location for those who value research for the sake of research. Its members do not have to teach students or give lecture courses. They are instead free to pursue research on whatever topics they find interesting. The permanent faculty receive good salaries to live in comfortable homes in a beautiful environment, spending the rest of their lives thinking about whatever they wish – the Institute does not face difficulties in hiring. It is permanently associated with Albert Einstein, who joined as one of the founding faculty members when he was at the peak of his fame and attempting to construct a unified theory.

In this attempt to construct a unified theory, Einstein was attempting to penetrate to the core of physics and to reveal its most basic principles. In his earlier discovery of general relativity, Einstein had done precisely this for space and time. Despite a level of experimental guidance that was somewhere between minimal and non-existent, he had been able to identify correctly the essence of the required theory and to write down the equations of general relativity. By obtaining a major result through the power of pure thought, Einstein became the exemplar of a theorist's theorist.

Einstein has now been dead for over half a century and physics since then has evolved dramatically. However his legacy, where the ideal result is one obtained without the need for experiment, persists. This style is characterised by a focus on clean problems, where the answers are well-defined and can be classified as either right or wrong. These problems do not have to be posed in the language of mathematics, but they should not be ambiguous, and formal elegance is appreciated in both the statement and the solution. The subjects addressed generally do not have direct empirical relevance; indeed, the question of relevance to observations can be seen as in slightly bad taste, betraying an overly vulgar interest in the quiddities of the world.

This style is well suited to the classic theoretical problems that cannot be addressed experimentally. We have seen some of these problems within this book. An example is the question, discussed in chapters 3 and 11, of what the entropy of a black hole corresponds to. As was said then, the entropy of a black hole is a measure of the number of possible ways of rearranging the innards of a black hole, and black holes have been known to have an entropy ever since the work of Stephen Hawking and Jacob Bekenstein in 1973. However, what exactly are these different possible inner configurations? Once you know what they are, how do you count them? This question is perfectly suited to this style of physics. It is well posed. It has a clear answer. Experiment will not help you.

In the context of string theory, this style is often associated with the attitude that the most pressing thing to do with string theory is to understand it. Applications come later: you cannot apply string theory until you first understand it, and the best way to understand string theory is in its own way and on its own terms. This means that the right calculations to do are those that reveal the inner world of the theory, often a world where there are ten dimensions of spacetime and a maximal amount of supersymmetry. Calculations, often with precisely chosen configurations of branes and in far more than three large spatial dimensions, are done to tease out subtle aspects of the theory.

On this approach, the mass of the electron, the value of the fine structure constant, the nature of the observed forces – these are accidental features of the world that belong to the same category of facts as the distance of the earth from the sun or the names of the emperors of Japan. It is unfair to level the charge that such calculations do not make predictions for observations, because that is not their purpose.

In the context of quantum field theory, this style appears in the study of quantum field theories that do not describe nature – but are amenable to exact solutions in the way that realistic quantum field theories are not. These theories have large amounts of symmetry, both super- and otherwise, and this large amount of symmetry can enable exact solutions to be obtained. These solutions are elegant, exact and geometric. They offer beautiful bridges into topics in modern mathematics and conceptual understanding of what makes quantum field theory tick. They are also applicable only to theories that are not applicable to this world.

For obvious reasons, this style of physics is never seen among experimenters. However it has always existed among theorists, with a strength that varies over time. It was certainly the dominant style in string theory throughout the 1980s and 1990s, rewarding with successful careers many people who have written a hundred papers or more, none of which concern either observations or data.

It is the style that was produced by those for whom Edward Witten is the model physicist, and who either consciously or unconsciously attempted to imitate him. Few of these, however, can match his combination of mathematical facility and physical insight, and indeed the work of Witten himself is not restricted to formal topics and has ranged widely though the field.

The style is certainly still common, although less fashionable. This is in part due to the prior harvest of low-lying fruit from the tree of theoretical ideas, and in part due to the coming of the Large Hadron Collider, which revitalised experimental particle physics.

In the late 1990s and early 2000s, following the rush of theoretical breakthroughs described in chapters 5 and 6, this style of approach to string theory was reasonably attractive. Times have changed since then, and it is certainly much harder now for a young person to be hired in a physics department if there is any implicit suspicion that they may regard experimental data as something beneath them.

12.5   IL FAUT CULTIVER NOTRE JARDIN

The English professional football leagues have four divisions.[8] At the top, there is the Premier League. This consists of twenty clubs, some of whom also compete in European competitions. The matches between these clubs are televised across the world, with the sale of the television rights bringing in enormous sums of money. The players in the Premier League earn huge salaries, taking home more in a week than an average worker does in a year. For their money, those who watch teams such as Arsenal or Chelsea or Manchester United get to see in the flesh some of the best players in the world competing at the highest level.

However, many of those who pay to watch live football do so outside the top division. Many people pay good money to watch not Liverpool FC but their near neighbours Tranmere Rovers. On a typical weekend, something like a quarter of a million people in England spend an afternoon watching football at a match outside the Premier League.

Why is this? The answer is not that the lower league clubs are merely affordable alternatives for when seats are unavailable for the big match. The same fans also went to Prenton Park rather than Anfield in the olden days of huge, cheap terraced stands, when even the biggest clubs rarely filled their grounds.

Nor is the answer the freefall of once great clubs down through the leagues. It is true that it is an unwritten rule of football supporting that an allegiance to a club, once given, can never be retracted. In this context there are few sins more heinous than changing club simply because the original choice now loses miserably and continuously, and one fancies watching graceful tiki-taka passing rather than up-and-under hoofball. Most lower league clubs have been so throughout their history, drifting like flotsam between second, third and fourth tiers via the occasional promotion or relegation. The choice to support Colchester United or Doncaster Rovers is not attended by any expectations of glory.

What are the positive reasons to watch lower league football? There are attractions among the lower divisions that are absent at the highest levels. As a supporter, you want your team to win every match, and you are disappointed when they lose. However many fans do not really at heart want their team competing at the top end of the top division. They prefer things the way they are. They are pleased that the club is part of the local community and does not arrange its preseason fixtures with an eye to the Shanghai fanbase. They like the fact that they know Bill on the turnstiles and Jenny who sells the

---

[8]As the name suggests, football is a game in which the feet and not the hands are used to propel the ball.

programmes by the main stand. They feel a connection with players who are local lads on a good wage, but who do not get to date pop stars. They enjoy recognising other supporters rather than being one of a gigantic horde. They like the fact that their club is a club and not a brand. They can feel that the club is *theirs*.

There are many benefits to life outside the headlights. There are also many scientists who genuinely prefer working in topics away from centre stage. They are happier when they know personally everyone else who works on a similar topic, and they enjoy getting together with them at small meetings to discuss progress. They like the ability to take a problem, think about it, turn it over in their minds, and make progress, away from the danger of being bulldozed by a stampede of hungry postdocs hunting the next big thing. Their work may not feature in press releases or in the plenary sessions of major conferences, but they also get to avoid the frustration of seeing others take over 'their' problem. They thrive best within a smaller ecosystem rather than seeking dominance.

The areas these scientists work on are not fashionable. Sometimes, areas are unfashionable for good reasons. The problems involved are difficult, and when solved confer no broader illumination. Questions that were once provocative become irrelevant as the subject advances – no one cares anymore about the fine details of the steady-state model of the universe. Those who continue working in such areas do fade away into gentle obscurity, accompanied by ever decreasing levels of interest in their research and ever increasing teaching requests from their head of department.

However, topics can also be unpopular for no better reason than that they are unpopular. Fashions are fickle. If an area is out of favour due to prejudice or misjudgement, those who are prepared to invest time in it will do well – and occasionally extraordinarily well. Both quantum field theory and string theory were once marginal areas ignored by the majority and populated by small numbers of scientists.

Some scientists enjoy hunting for such undeservedly unloved areas and cherishing them. The future for those in such areas is bright. Even so, once success comes there are different responses to it. The most natural attitude is to continue working on the same topic, and as a founder of the field be carried upwards on a rising tide of interest. For some though, part of the pleasure of working on an unfashionable area is the opportunity to think deep, foundational thoughts and thereby shape its development. As the topic becomes part of the mainstream, and so enters the province of the many, their job is done – and they head off to search for another neglected patch in order to make it bloom.

This pattern is seen in many of the very best scientists, whose skills are not fully tested by repetitive calculations. They are possessed of originality of approach, the technical ability to change topics, and the confidence that having done so they can find insights that others have missed. These skills allow them to contribute in many different areas. As one example, as a graduate student Frank Wilczek did work on the strong force that would win him the

Nobel Prize. However rather than simply refine predictions for the behaviour of the strong force, over the next decade he also made central contributions to the theory of axions – the type of particle we met in chapter 10 – and also discovered a novel form of quantum statistics, called anyonic statistics, which applies in two spatial dimensions and is relevant to many problems in condensed matter physics.

This chapter has described some of the many different ways to be a theorist. There are even more ways to be a scientist in general. While all types are needed for the subject to flourish, I now turn instead to arguments that string theory is not a flourishing subject, but is rather a pestilent weed in the garden of science.

CHAPTER 13

#EpicFail? Criticisms of String Theory

Toddlers and Taleban alike know that it is always easier to knock down than to build up. Criticism is cheap and criticism – of anything – is easy. The critic gets to pick his point and time of attack. The critic does not have to be fair, and the critic does not have to provide either a solution to the issues he raises or any alternative proposal to deal with them.

For a topic whose subject matter is at almost maximal remove from daily concerns, string theory attracts a surprising amount of emotion. At the time of writing, the first three options on the google autocomplete of 'String theory is' are 'dead', 'wrong' and 'bullshit'. It is a topic on which people have real and polarised opinions. Having last encountered any science or mathematics amidst the fraught years of adolescence is no obstacle at all to holding strong opinions on string theory's relative merits compared to other ideas for quantum gravity.[1]

Of course, criticism has many positive features. The concept of the loyal opposition is a glorious feature of parliamentary democracy. The good faith exchange of contradictory views can sharpen vague ideas into blades that cut. False pretensions and claims are blown away by a need to provide clear answers to clear questions. Critical debate can bring the point of disagreement into focus, thereby showing where hidden assumptions are entering in. Every scientist grumbles about the referee reports they get back on the papers they submit for publication, but few would deny that on the whole these reports lead to improvements in the quality and readability of the papers.

There are many criticisms that have been levelled at string theory, and the aim of this chapter is to provide responses to these criticisms. In doing so

---

[1]This passion for abstract topics cannot however compete with fourth-century Byzantium under the emperor Theodosius, when fishmonger and carpenter would passionately debate in the marketplace the relative merits of the homoi-ousian and the homo-ousian nature of Christ.

I have tried to be fair and to do my best by arguments even when I disagree with them. In my formulations, I have attempted to capture the spirit of these criticisms and to put them in their most convincing form. Sometimes they involve unspoken assumptions. Whenever possible I have sought to allow these assumptions and to meet the criticisms on their own ground.

However, I also want to make clear that what I address in this chapter are criticisms of string theory, and not the promotion of any individual alternative theory. This chapter deals with the arguments made that string theory is wrong or misguided, and not with arguments that some other theory is right. This is a case to be made by those who believe in it, and I have provided some references in the bibliography for those who want to pursue their arguments. For this reason this chapter will not deal with any criticisms of the form 'But string theory is so much worse at frying burgers than my theory'.

I will give partial consideration to such arguments in the next chapter, which is the counterpoint to this one. The next chapter deals not with why string theory is wrong, but with why string theory is right. It gives the positive case for why string theory has been so much more successful than any other proposed theory of quantum gravity. In doing so, it will in part address these unfavourable comparisons of string theory to other theories, as well as including some brief comments on these proposals. These comments will be brief – this book is not primarily about quantum gravity and it is certainly not about all theories of quantum gravity ever proposed.

So, what are the reasons put forward that string theory is *ex operibus diaboli*?

> CRITICISM: The attractiveness of string theory comes from its claim to solve the high-energy (sometimes called ultraviolet) problems of supergravity by making the divergences finite. However, there is no actual proof that string theory is finite. The calculations only hold at the lowest orders in perturbation theory and have not been extended further. Beyond these, finiteness is simply a conjecture, but not one that has actually been proven. As such, the finiteness of string theory might be an interesting idea, but one should not place too more store by it.

The strength and weakness of this point lies entirely in the word 'proof'. 'Proof' is a heavily loaded word. It is also a word that has more in common with mathematics than physics. The interesting structures of physics are too complicated to be described at the level of mathematical detail that can happily accommodate the mathematicians' notion of proof.

In 1984, one of the main selling points of string theory was that it offered a possible answer to apparently insuperable problems with the supergravity theories. This answer was partly conceptual and partly calculational. There were indeed calculations that gave finite answers where supergravity gave infinite answers. However, and at least as importantly, there were also conceptual arguments, through which the extended nature of the string provided a reason why these problems should be absent in string theory.

One aspect of this argument was that strings are extended objects and so tend to smear out infinities associated to point particles with no spatial extent. Another more technical point was that the structure of string theory offered a way to reinterpret any high energy problems of supergravity as low energy questions. On this way of thinking, the short-distance infinities of supergravity could be re-understood as long-distance effects. However, divergences associated to long-distance effects were already well understood through studies of quantum field theory and were known to be harmless. These arguments made string theory attractive in 1984: it proposed a new way to solve an old problem, and wherever the new ideas could be tested, they worked.

At that time, the question of whether all these ideas really did work as they appeared was a good and interesting question. Were the cancellations that had been found in the superstring merely a lucky coincidence? The arguments for finiteness worked neatly in the simplest settings – at the lowest orders in perturbation theory. However, even if they were saying something, there could still be something more that had been missed. Was superstring theory really consistent? This is ultimately what the criticism above asks – is string theory actually a theory that makes sense?

Superstring theory in 1984 was a relatively new and poorly understood structure focussed on the particular problem of quantum gravity. In this context, any proof of finiteness would have been very welcome. Such a proof would have, by necessity, automatically greatly extended the technical tools available in string theory. The techniques then used for describing fermions became prohibitively complicated beyond the lowest orders of perturbation theory, and of necessity any proof would have had to include methods for working at all orders in perturbation theory. Any general proof would have greatly extended the relatively few calculations that existed and would have offered clues to how the theory should be developed.

However, by the time we reach 2015 string theory has produced so much more of interest that this question of 'proof' is far less interesting. The number of positive, surprising and correct results produced by string theory is now so large that there can be no reasonable doubt that string theory as known today does represent a consistent mathematical something, even if it is not possible to define exactly what that something is. As one example, we saw in chapter 8 one of the highly intricate formulae that string theory reproduces in the AdS/CFT duality. It is beyond reasonable doubt that these agreements are not simply a fluke.

The technical argument for the finiteness of string theory is that the structure of the theory always allows potentially dangerous high-energy divergences to be reinterpreted as harmless low-energy divergences. In any place where it can be tested – including for far more complicated perturbative calculations than could be performed in 1985 – this principle has continued to hold over the last thirty years.

We still do not know fully how string theory works or what its most fundamental principles are. However, it by now requires something approaching dishonesty for a professional to doubt that it exists as a consistent theory of something. A Victorian engineer confronted with the latest mobile phone would be totally baffled as to how it works. He would have no possible conception at all as to the nature of the internal circuitry – the transistor would not be invented until long after his death – but he would also have no doubt that this circuitry worked.

It may still be said: that may be so, but why cannot someone still just take a few months to write down a proof? The answer is that physics is not easily amenable to proofs, and proofs cannot be found even for topics far simpler than string theory. As mentioned in chapter 8, there is a one million dollar prize available from the Clay Mathematics Foundation for proving one of the basic features of the Standard Model: the presence of a mass gap in the strong force. This is the statement that there are no massless particles charged under the strong force – there is a 'gap' to the first allowed mass. Compared to questions involving quantum gravity, this is a baby problem. The Standard Model is much simpler than quantum gravity. The techniques are far more understood. This question is also accessible to experimental study. There is also a *one million dollar incentive* – and yet there is still no proof.

> CRITICISM: String theory comes in so many forms that it is impossible to make any predictions. There are an almost infinite number of ways to compactify down from ten dimensions to four. Each way represents a different string theory, and each will lead to entirely different physics. String theorists themselves say that there are 10^{500} such possibilities, and so if you can get 10^{500} different theories you can get anything you want out. A theory that can predict anything is a theory that predicts nothing. A theory that makes no predictions and is not falsifiable is not science.

This criticism contains several errors and exaggerations, which I will address below. The criticism also contains an attitude to falsifiability characteristic of Popperians of the strict observance, which I note but shall not challenge.

The first main error is that it conflates the questions of 'What is science?' and 'What is the state of current technology?'. It is clear that the ability to test any idea experimentally is a function of the technology of the time. Nuclear physics was just as true in the stone age as it is today,[2] and it will remain just as true if we are returned thither through some war or catastrophe. Today, nuclear physics and the associated quantum mechanics is testable. In the past they were not, and in the future they may not be either. Their scientific truth, however, endures.

---

[2] And earlier – around two billion years ago, a natural nuclear reactor operated at Oklo in Gabon in central Africa, fissioning much of the uranium present via a chain reaction.

That said, it is always better – not least for the scientists involved! – when ideas can be tested within a few years of their proposal, or at most within the lifetimes of the scientists. It would have taken the heart of a Vulcan not to rejoice in Peter Higgs's pleasure in living to see the discovery of the Higgs boson in 2012, at the age of eighty-three and almost fifty years after his paper on the topic. No one enters science for the money, but that does not confer immunity from the human desires for recognition and acclaim.

Science is also healthiest when the interchange between theory and experiment is rapid. Wrong ideas, like aggressive weeds, are best killed quickly, and experiment is the best killer of them. Science moves fastest when theoretical ideas are closely coupled to experiment. However, *sub specie aeternitatis* it is ultimately irrelevant whether bridging the technological gap required to test a theory takes ten years, a hundred years – or longer. Democritus was no less right that the world is made from atoms for having died over two thousand years before the construction of the periodic table.

It is clear that the natural scale of string theory is not the scale of atoms and is not the scale of the Large Hadron Collider. It is the scale of quantum gravity, and whatever that may be precisely, we certainly know it is far smaller than any distance scale we can currently access. Our inability to access this scale is technological, but not a question of principle. Given magnets large enough and long enough, we know how to accelerate protons to quantum gravity energies.

However, the Large Hadron Collider currently represents the best that we can do. If money were no object, we could do better; but as seen in chapter 11 even then there is no open path to studying physics directly at the Planck scale. All current technologies fail long before we reach these scales. While history teaches us to be exceedingly modest when attempting to constrain future ingenuity, it is clear that predictions for the Planck scale are for the moment a question of principle rather than practice.

Nonetheless, what are the predictions of string theory at these quantum gravity scales? In brief, they are extra dimensions, extended objects and soft scattering. As we have seen in chapter 10, from a four-dimensional perspective extra dimensions manifest themselves as additional particles: ten-dimensional gravity has many more internal degrees of freedom than four-dimensional gravity. This statement remains true whether the extra dimensions are classical geometric dimensions or quantum stringy dimensions with no easy classical interpretation.

Likewise, strings are characterised by an enormously rapid – an exponentially rapid – growth of the number of harmonics with energy, corresponding to the many possible directions in which a string can vibrate. As we have also seen, the scattering of strings (or any other extended object) at high energy furthermore has the distinctive feature of soft scattering – colliding objects have minimal tendency to go off at right angles from the collision axis.

These predictions are not hard to test. Once you have a microscope that is capable of resolving sufficiently small lengths, there is no mystery about how to test the relative claims that the electron is a particle or the electron is a string. You use the microscope, and you go and look. Indeed, no philosophical agonising about falsifiability occurred when string theory in its original incarnation was proposed as a theory of the strong force, and the characteristic length of strings was thought to be a femtometre. The reason string theory was originally ruled out as an account of the strong force was precisely because, as more experimental data arrived, its predictions totally and spectacularly failed to accord with this data.

If you can look at the quantum gravity scale, string theory is then not hard to test. At this point a rider is sometimes added to this objection: what about M-theory? The different string theories are all meant to be different limits of M-theory, but the equations of M-theory are unknown. If you cannot say fully what string theory really *is*, how can you say it is testable? How can you make any statements about predictions without a full definition of what is meant by the theory?

There are two answers to this. The first, conservative, one is to say simply that the above statements about testability apply only to all the work done on string theory in the last thirty years. In that string theory is a topic that has absorbed real people's time, it is testable in this sense, and these statements certainly apply to all the work that caused anyone to be interested in string theory in the first case.

The stronger, but still reasonable, response is that extra dimensions and extended objects are always present in string theory, and extra dimensions do carry physical meaning. As we go to higher energies, these extra dimensions become apparent and the number of particle-like states grow enormously. Likewise, the presence of spatially extended objects – whether strings or the branes of M-theory – is something that one can always look for once sufficiently high energies are attained.

Returning to the original criticism, the second inaccuracy is that it is not true that an almost infinite number of ways to go from ten dimensions to four dimensions implies an almost infinite number of possibilities for four dimensional physics. The number 10^{500} sounds large – and it is. However, as discussed in chapter 6 it is dwarfed by the number of genetic permutations that can arise when mummy and daddy get jiggy and make a unique human being. Despite this, observation of siblings and their parents belie the notion that infinite variation is therefore possible – and while there may be far more than 10^{500} possible human genomes, we can predict with good accuracy the number of fingers someone has.

Specifically, we have seen in chapter 10 that theories with extra dimensions leave characteristic legacies in lower dimensions. Almost always, there are additional light particles with very weak interactions: moduli, additional hidden forces or axion-like particles. These particles are simply a feature of extra dimensions and are present in any theory with extra dimensions. Their existence is therefore insensitive to the many different ways of moving from ten to four dimensions. It is not a theorem, but I am trying and failing to think of any counterexamples.

As we have also seen in chapter 10, there are many ways to look for such particles experimentally. These searches are not easy, and success is not guaranteed, but this situation is hardly unique to string theory.

It is certainly true that, as a fundamental theory of nature, string theory is hard to test. Of course, it would be undeniably nice to have an experiment with existing technology that was capable of giving a definitive answer about whether string theory – or anything else – was a correct description of physics at scales fifteen orders of magnitude smaller than those we are able to probe directly. To which the only response is: yes, it would be nice.

> CRITICISM: Modern physics, of which string theory is an example, ignores philosophy and does so at its peril. It is not reflective, but instead attempts to develop the subject following the 'shut up and calculate' tradition. In doing so it cuts off the hand that feeds it; it believes it can answer foundational questions while ignoring foundational thinking. The development of relativity required input of philosophical ideas such as Mach's principle; there is no reason to suppose the much harder problem of quantum gravity should be any different.

The essence of this criticism is that many of the deepest problems in physics are philosophical in nature. What is the nature of space? What is the nature of time? What are the basic principles that any quantum theory of gravity must satisfy? The argument made is that blind calculation is not enough – these questions cannot be answered without philosophical reflection, and that this process has been systematically rejected. The particle physicists of the 1960s and 1970s, flush with data, could get away with rejecting philosophy. However for problems without abundant data, this attitude is presumptuous at best and idiotic at worse.

Where this objection chiefly fails is in a conflation between the concept of 'philosophy' and 'what those calling themselves philosophers do in the philosophy department'. Nature does not divide itself by university department. Up until the nineteenth century, what we now call science used to be called natural philosophy. Isaac Newton's most famous work is called '*Mathematical Principles of Natural Philosophy*'. In the title, he makes the statement that natural philosophy is best done with the language of mathematics – while also gently alluding to Descartes' non-mathematical 1644 work *Principles of Philosophy*. 'Science' at that time was just natural philosophy – the philosophy of nature.

While the name has changed, the essence of the subject has not. For example, Richard Feynman was famously disparaging about philosophy – 'low level baloney' was one of his more polite comments. But, Feynman was also the person who reformulated quantum mechanics as a sum over all possible histories of a system. If you want to know what is the quantum mechanical probability for a particle to go from A to B, Feynman said, then you can do it by adding up contributions from all the possible paths there are from A to B.[3] Which way did the particle go? It went every which way. All paths contribute, and we do not and cannot say more. This is a deep truth about nature, and it a deep truth that deals with the same branch of knowledge that Aristotle's *Physics* did.

An example more relevant to string theory is the case of the holographic principle. This is the statement that the physics of a gravitational system in D dimensions can be captured by the physics of a non-gravitational system in $(D-1)$ dimensions. This is a statement that is made sharp in the AdS/CFT correspondence, which gives a precise mathematical formulation of it. This is one of the major components of string theory research in the last twenty years, and no criticism of string theory can simply excise this topic from consideration. But – how can the holographic principle *not* be regarded as philosophy? In any way that philosophy is worthy of the name, how can such a deep statement about nature not be called philosophical? It is every bit as deep as any of the ideas that fed into the development of relativity, and the sharpness of the calculational tests of it can only be a virtue and not a vice.

AdS/CFT is an example of a duality. There are other dualities that provide similar examples. In chapter 5 we encountered T-duality, which is essentially the statement that in string theory very small spaces are indistinguishable from very large spaces. Despite its surprising identification between two very different geometries, T-duality is still one of the best understood dualities in string theory. The mathematical subject of mirror symmetry that we encountered in chapter 9 can be seen as a generalisation of it. How can T-duality not be regarded as a philosophical statement about what space really is? Furthermore, it is a result backed by precise calculations. Just as in the time of Newton, a statement should not be seen as less philosophical merely because it is backed by mathematical evidence.

My general response to this criticism is then that on any historic reading of what counts as philosophy, or on any self-respecting notion of what philosophy encompasses, string theory does not ignore philosophy. It is instead part of (natural) philosophy.

The narrower statement that string theory is deficient as a theory of quantum gravity because it pays insufficient attention to what is going on in the philosophy department is simply weak (or, as Feynman might say, baloney). It is the same sort of baloney as the argument that your plumber might not be able to fix the drains because she is not an expert on the Victorian novel. It may be true that a plumber would be a better plumber for a wider knowledge of literature, but it is hardly the crucial aspect of the job. One part of my employment involves teaching physics at New College in Oxford, and one of the many pleasures of working in an Oxford college is a greater-than-average

---

[3]More precisely, probability in quantum mechanics is the square of an amplitude. Feynman's prescription was to first sum the individual amplitudes for each path and then to square this sum.

exposure to professional philosophers – who are intelligent and sensible people who do not make this sort of silly argument.

> CRITICISM: String theory is too mathematical and has lost touch with actual physics. Physics advances through experiment, and it is extremely dangerous to believe it is possible just to think one's way to the answer without any input from observation. Practitioners of string theory have become obsessed with mathematical beauty and regard it as a reliable guide to truth. However their idea of 'beauty' may be false, and other people may find different ideas beautiful. Furthermore, it is not mathematical beauty that is relevant in evaluating a physical theory, but success in explaining experimental data. The 'beauty' beloved of string theorists leads them to ten spacetime dimensions; this is in manifest contradiction with observation.

This criticism certainly contains elements of truth. There are many who work on string theory who are entirely uninterested in either observational input or output. It is not what motivates them. They are interested in the formal structure of theories or in mathematical applications of them. The prospect of explaining experimental data is not what gets them out of bed in the morning.

There is nothing wrong with this. Mathematics is a worthy subject, and it is not less important because it does not involve experiment. Studying string theory for its mathematical applications is an entirely sensible reason to study it. There is a valid question, which I have sympathy with, as to whether *too many* people are currently working on the subject for reasons only tangentially related to physics. This is a legitimate question about distribution of funding, effort and resources, but it is a question of a different kind.

What is not defensible is the idea that string theory cannot be relevant for physics *because* many aspects of it involve advanced mathematics – where 'advanced' means significantly more mathematics than was needed for the formulation of either the Standard Model or general relativity. Mathematics is certainly not the only guide to truth, but it is historically true that advances in mathematics and advances in physics have fitted together hand in glove.

Furthermore, what precisely is meant by 'too mathematical'? Difficult mathematics has been encountered in physics before. This is what Max Born, one of the founders of quantum mechanics and winner of the 1954 Nobel Prize, had to say about the start of quantum mechanics:

> By observation of known examples solved by guess-work [Heisenberg] found this rule and applied it successfully to simple examples ...
> I could not take my mind off Heisenberg's multiplication rule, and after a week of intensive thought and trial I suddenly remembered an algebraic theory which I had learned from my teacher,

Professor Rosanes, in Breslau. Such square arrays are well known to mathematicians and, in conjunction with a specific rule for multiplication, are called matrices.

Matrices are now the type of diddy topic that are taught in school and professional physicists end up unable to remember not knowing. Looking further back, Cartesian coordinates – labelling graphs with an x and a y axis – were also at their time a shocking innovation. However useful, obvious and natural they may seem to us, the hard truth is that their discovery eluded Greek, Arabic, Chinese and mediaeval science and mathematics.

The problem with 'too much mathematics' as an objection is that it appears to be shorthand for 'there is too much unfamiliar mathematics compared to the mathematics I learnt as a student'. It is clear from history that advances in physics have very often required mathematics that was unfamiliar and that initially appeared bizarre. Mathematics that is necessary becomes familiar, and mathematics that becomes familiar becomes easier.

The complaint that whatever progress has been made in string theory has not been through explaining experimental data is a true one. It is also a slightly unfair one. The book started with an account of the unreasonable success of the Standard Model, a theory that has been far more successful in explaining experimental data than it ever deserved to be. All data in particle physics is consistent with the Standard Model. So far, all searches for qualitatively new physics have been without success – and if anything is to 'blame' for this fact, it is the laws of nature.

Null results do give (some) information, but they are nowhere near as informative as discoveries. It is almost a tautology that if there is any progress that can be currently made about physics at quantum gravity scales, this progress will require more than just experimental data – and mathematics will play some role in it.

CRITICISM: One of the major features of Einstein's theory of general relativity is that it is background independent. Its formulation does not depend on a choice of coordinates. All that really exists are relations between objects, and any fundamental formulation of physics must be done in a way that does not depend on any particular choice of coordinates. In particular, any correct theory of quantum gravity must be background independent.

However, string theory is not background independent. The standard formulation of string theory is in terms of an expansion in terms of small perturbations about a particular spatial background. String theory therefore always depends on a choice of a background. Its physics is not background independent, and consequently string theory is not a theory of quantum gravity.

To my mind, the problem with this view is that it is based on a rather fixed ideological belief concerning what quantum gravity *must* be. The criticism is founded on the notion that one can first guess or deduce the principles underlying fundamental physics, and then construct the theory according to the principles. It expresses an overconfidence in the ability to know how everything will turn out, independent of input from either experiment or calculation. I am reminded of the (possibly apocryphal) response of Niels Bohr to Albert Einstein when he expressed his doubts about quantum mechanics:

> Einstein: God does not play dice.
> Bohr: Don't tell God what to do!

Let me make two more detailed responses. First, geometry is not fixed but manifestly dynamical in string theory. The fields that describe spacetime are not static. They have equations of motion, and these equations of motion cause them to change. In that string theory is an expansion about a fixed background, it is also a background that changes dynamically according to Einstein's equations. Small changes build up to large changes, and large changes can be as large as one wishes.

There is also an important distinction between the statement that the physics must be background independent, and the statement that the formulation of the physics must be background dependent. This may seem unclear. While quantum gravity may be esoteric, there is a more familiar topic in which one can re-express this same issue: cartography and the making of maps.

How do we describe the geometry of the earth? There are two ways, one 'background independent' and one 'background dependent'. The background dependent way is through an atlas of charts. If you purchase an atlas, on every page you will find a map of a different part of the earth. Depending on the purpose of the atlas, these maps can be of varied quality with varying levels of detail. They contain cities, towns and villages. They contains the contours of the land and the depths of the sea. They contain ship wrecks and sandbanks, castles and churches. Each chart only describes a small part of the overall picture: a ship sailing to Archangel will have little use for a map of Cape Horn. The charts also depend entirely on coordinates, as they have latitude and longitude lines stretched across them. Patched together however, the charts describe the surface of the entire earth: they are good for any purpose.

There is also a 'background independent' way of describing the earth. This is through a globe. A globe provides a visualisation of the full geometry of the earth. With a globe there is no requirement of labels for latitude and longitude, or indeed any other choice of coordinates. Globes preserve perspective and area in a way that is not possible to do with an atlas, and they are excellent educational tools.[4] However – it is not reasonable to argue that globes are 'right' and atlases are 'wrong'. An atlas – which uses particular choices and

---

[4]Maps in an atlas can either preserve area and violate angles, or preserve angles and violate areas. The most familiar atlas projection of the globe is the Mercator projection. This preserves angles but does not preserve area – making the relative size of Europe compared to Africa appear far larger than it actually is.

charts to label every part of the globe – is a precise and correct description of the globe. In fact, the mathematical definition of any geometric space, technically called a *manifold*, is done precisely in terms of an atlas of charts. So there is nothing 'wrong' with the choice of coordinates – it is the choice to use an atlas of charts rather than a globe.

The second objection is that 'background independent' is a slogan, and a hollow slogan without deep content unless further accompanied by a notion of what *a* background is. To claim to be independent of all backgrounds, it is first necessary to say what these individual backgrounds actually are.

The simplest possible background is flat spacetime: a background that was already present in special relativity. Slightly more complicated backgrounds are the curved but classical geometries that arise in general relativity. These solutions are still, however, well approximated by (generalisations of) Einstein's theory of gravity.

What about more complex backgrounds than these classical, weakly curved spaces? In string theory, it has required many centuries of work to determine the large variety of possible different backgrounds that are permitted in the subject, and 'background' in string theory is a much richer concept than in general relativity. As seen in chapter 11, it must enlarge to include geometries that are of different topology. 'Topology' refers to the properties of objects that remain unaltered under any smooth change. The shape of a bagel cannot be deformed into the shape of a tennis ball no matter how much you knead it – you have to tear it. Any such geometric transition is entirely impossible in Einstein's theory of gravity, as you cannot tear space. As also seen in chapter 11, string theory contains controlled examples in which the space changes topology. You can smoothly change the topology of spacetime in string theory without anything funny happening.

Secondly, the concept of a background must also include examples where the background smoothly deforms, in a calculable fashion, from the classical picture of Einstein into a form of 'quantum geometry'. In quantum geometry the background no longer admits an interpretation in terms of classical notions of space. Coordinates no longer have any meaning – these represent an idea that sensibly applies only for the classical geometries your grandparents grew up with.

As a final illustration, the range of backgrounds must also include geometric spaces of different dimensionality. As we saw in chapter 5, one of the most surprising results from the mid-1990s was that string theory taken as a whole has limits in which it is either a ten-dimensional theory *or* an eleven-dimensional theory – and it is possible to interpolate between the two.

While 'background' in string theory is mostly a richer concept than in general relativity, it is also in some ways poorer. There are backgrounds that look very different in a classical theory of gravity, but that are identical in string theory. In string theory, T-duality implies very big spaces and very small spaces are the same. As backgrounds, they are absolutely identical. They are one and the same. This is not at all obvious at the outset, and it can only be seen by looking at the actual equations of string theory on an actual background.

All these results were found through hard calculation, by looking at particular backgrounds in detail and understanding what happens as small changes are made near those backgrounds. None of these results would have been easy to guess in advance.

The danger with the assertion that background independence is a guiding principle of quantum gravity is that it tends towards an impoverished view of what is possible. Real content comes from knowing what all the possible 'backgrounds' can be. Once you know what all the possible backgrounds can be, you are a long way towards knowing what quantum gravity is.

What the criticism does correctly capture is the fact that in an ideal world, you would have a formulation of string theory that gave you a view from which all these surprising results become 'obvious'. From the right perspective, crazy relationships just become simple consequences of general principles. When this is attained, it produces one of these glorious moments of scientific ecstasy when understanding brushes aside confusion.

This perspective does not yet fully exist for string theory. However the fact that such a formulation does not yet fully exist does not make string theory wrong – it just makes it a topic of research. It is like saying that because no-one can yet prove the elementary (and apparently true) statement that all even numbers can be expressed as the sum of two primes, we have no theory of prime numbers.

In the end, 'background independence' is a rallying call. If someone believes that the only way to make progress is by following this principle and writing down, in one go, the full theory of quantum gravity, then that is what they believe, and no amount of result or calculation will convince them otherwise. In this respect, extensive argument with proponents of this view becomes like a discussion with either committed Marxists or members of the Chicago school of economics, where independent of the question the answer is either 'dialectical materialism' or 'monetarism and free trade'.

> CRITICISM: String theory receives too high a fraction of the available resources for fundamental physics. String theory has promise, and it is reasonable that some people are interested in the subject and work on it. However, there are many approaches to quantum gravity and this should not be to the exclusion of other methods. In the same way that retirement savings should not be entirely invested in a single stock, resources in physics should be far more equitably distributed so that similar levels of attention can be paid to each of the different approaches to quantum gravity.

While it superficially sounds entirely reasonable, this criticism contains two implicit assumptions. First, it assumes that there is actually a lot of money spent on string theory. Second, it assumes that this money comes from a large pot of soft goodies, which is jealously guarded to prevent it being shared out. In this world, there should be more than enough money to go round, with some to spare. The only reason it is not is because of bad behaviour by string theorists, who hoard these resources, keeping them for themselves and their friends.

The fault with this criticism is that it supposes an idyllic world entirely disconnected from the practicalities of funding. In the real world, scientists almost entirely get money to do research by asking funding bodies for grants. If I want to get money to do research on string theory (as I do, and as I have done), I do not do so by asking a committee full of my chums. Instead, I have to make my case to a panel from many different specialisms. The vast majority of this panel will have never worked on anything even tangentially related to either string theory or quantum gravity, and indeed may not even be working on particle physics. I have to convince this panel both that I, as an individual, am worth funding and also that the topic I propose to work on deserves public money. For the largest grant, by cash terms, that I have been awarded, the relevant committee involved sixteen people, of whom a total of one – precisely one – was in even the broadest and most generous interpretation of 'my area'.

Success in such grant applications depends on both tangible and intangible factors. The tangible factors involve both past history and publication record: the papers you have written and the number of times they have been cited. Past performance is no guarantee of future success, but it certainly helps in a grant application. There are also the intangibles – the fluency of a presentation and the ability to make a research proposal convincing and comprehensible to those outside the subject, all mixed with the individual perversities and predilections of the interview panel and its members.

The people making the decisions to spend money on string theory, then, are not string theorists. Grants are hard to get and grant applications are competitive. The success rate for the major grants that launch independent careers can be smaller than ten per cent, and the large majority of applications fail. Scientific funding is not a Care Bears' tea party. Panels have to decide where limited resources can be most productively spent, and every penny obtained is obtained by convincing those outside your field that what you do is worthwhile and deserves to be funded. There is no soft pot of money available for those who would like a good salary to develop their own theory of quantum gravity at the taxpayers' expense.

Why has string theory been successful in this endeavour? The long answer has been given throughout this book. The short answer is that, as seen in chapters 8 to 11, string theory has proven to be so much more than just quantum gravity – and by doing so it has become attractive to large numbers of scientists.

In the next and final chapter I summarise this positive case for string theory, and also explain why string theory has in fact been preferred over other alternative theories of quantum gravity.

CHAPTER 14

Why String Theory?

14.1 REASONS FOR SUCCESS

Why have so many chosen to study string theory? It is an esoteric theory that requires many years of study to appreciate. It is not a commercial subject offering large financial rewards at the end. It is also not experimentally validated. Unlike the Standard Model, which has run the gauntlet of data so many times, there is no direct evidence that string theory is a correct theory of nature.

What is the reason for the success of string theory as an idea? As we saw in the previous chapter, there is a negative account of this success which goes something like the following. String theory is an approach to quantum gravity, one of many. All have virtues and all have problems. As it is extraordinarily difficult to find direct experimental probes of quantum gravity, we cannot know which of these many approaches is the correct one. Given this uncertainty, it would be best to pursue research in quantum gravity through a happily diverse community working on a variety of areas. Instead, powerful people working on string theory are only willing to hire other string theorists. These people in turn then hire more string theorists, leading to a situation where the only route to a professional career is through working on string theory. String theory then became the dominant approach to quantum gravity not through any merit of its own, but rather through the Cosa Nostra method of protecting family and destroying competitors.

While this argument is basically false, it does contain a grain of truth. The grain of truth is that it is simply true that the quality standard required for the student of Professor Bigshot Largecheese to get a job is lower than that required for the student of Professor Podunk Smalltown. Senior people with power and influence do try and lean jobs towards their friends and connections, and this certainly exists as a factor in determining who gets hired to which jobs at which universities. While this nepotistic habit of favouring friends and mentees is not to be commended, most would also regard the suggestion that this illness is particular to the groves of academe as reflecting a touching naivety about the way the world works.

What then lies behind the success of string theory? A wise and wealthy entrepreneur once observed that the route to riches in a gold rush is not through discovering nuggets but instead through selling shovels and pans. In this analogy, the large nugget of pure gold corresponds to the true and experimentally validated theory of quantum gravity. The shovels and pans are calculational techniques, mathematical insights and applications to other parts of theoretical physics.

One of the major themes of this book has been that while there are many people who can loosely be called string theorists, very few of them actually work on quantum gravity or have their main interests in quantum gravity. The above complaint is that nefarious trickery lies behind the large numbers of people working on string-theoretic approaches to quantum gravity. The more accurate statement is that string theory has been found useful by many people who are in no way interested in quantum gravity. The enormous growth and professional success of string theory is because so many physicists with no *a priori* interest in it found that it had interesting things to say about topics that they cared about.

One example of this is the community of those interested in higher-dimensional theories of gravity, and in particular in the supergravity theories with ten or eleven dimensions. Today, these subjects are interwoven with string theory. Starting graduate students confidently describe ten-dimensional supergravity as the low-energy classical limit of string theory, and eleven dimensional supergravity as a classical limit of M-theory. As seen in chapter 5, it was not always so. These subjects have different parentage. Interest in higher-dimensional gravity theories came from the development of classical gravity theories, by a community founded on intricate and difficult computations in supersymmetric theories of gravity. During the 1970s and 1980s they developed these theories of higher-dimensional supergravities and studied many of their properties.

However following the explosion of interest in string theory in 1984, the validity of supergravity came under attack in many ways. Superstring theory showed that the right number of dimensions was ten – what were these people doing studying theories with eleven dimensions? These theories did not exist in quantum gravity. Even ten-dimensional supergravity was only an *approximation* to string theory. It was not the real deal. Why study the approximation when there was no need to approximate? Instead, it was better to study string theory – these classical computations could have nothing to say about the consistent quantum mechanical theory that was string theory. Instead of studying one-dimensional extended objects – strings – the supergravity folks were studying branes, objects with several spatial dimensions. This was clear further evidence, it was thought, that they did not know what they were doing!

As also discussed in chapter 5, this situation all changed in the middle of the 1990s, as it was realised that the branes that were being studied were an integral part of string theory – and furthermore had a remarkably simple description within string theory that greatly clarified their properties. The study of branes in supergravity and the study of branes in string theory were different sides of the same coin. They were part of the same subject. From the outside, the community of people studying branes in supergravity became 'string theorists' – but this occurred not through conversion but rather from a natural merger of the two subjects. If you wanted to understand non-perturbative effects in string theory, it would have been silly to shun the gift horse saddled with diamonds and rubies that was the supergravity literature on branes and eleven dimensions – and in a similar way the string theory picture of D-branes provided a new, simpler perspective on branes in supergravity.

Another example is the case of quantum field theory. Quantum field theory is the great foundational concept of the Standard Model. As seen in chapter 8, the wake of this success contains many groups of scientists who work on understanding quantum field theory better. How does quantum field theory behave when all the interactions are very strong? How does quantum field theory behave when there are lots of additional symmetries? Are there simple, particularly symmetric versions of quantum field theory that can be solved outright? What universal rules can be found for quantum field theory? This busy hive of scholars were and are only tangentially concerned with matching theoretical predictions to experimental data; they instead sought a better understanding of a general type of theory that was used in our accounts of nature. This busy hive of scholars also had no interest in quantum gravity and theories thereof: they wanted to understand quantum field theory in four dimensions and not quantum gravity in ten.

Nowadays, these quantum field theory experts are also experts about string theory. They discuss knowledgeably about higher-dimensional gravity and the different solutions of general relativity in both five and ten dimensions. They talk enthusiastically about strings and branes, and strings ending on branes, and branes intersecting other branes, and strings stretching between branes and the full general theory of branes. They go to string theory conferences, talk about string theory and get called string theorists. However, this is not because they were zombified into the Great Undead String Army. It is not that they came to string theory; string theory came to them and explained that many interesting properties of four-dimensional quantum field theories are best understood via higher-dimensional theories of gravity. As described in chapter 8, this arose through the shocking duality that is the gauge-gravity correspondence – in certain limits, higher-dimensional theories of gravity are *one and the same* as lower-dimensional quantum field theories.

As we saw there, 'dictionaries' were produced that translated between quantities in higher-dimensional gravity theories and quantities in lower-dimensional field theories. It is said that in transnational commerce you can buy from a company speaking your own language, but to sell to them you must speak their language. This is how the influence of string theory has spread – it came to the field theorists and showed them how it could be used to compute quantities of interest to them. As string theory was used to solve problems that were purely problems in quantum field theory and defined only within that framework, string theory grew to encompass the large community of quantum field theorists. This was not because field theorists suddenly became interested in string theory as a solution to quantum gravity, but because string theory was able to show them a way to solve their own problems and on their own terms.

The position of string theory within mathematics has followed a similar path. Various areas of mathematics have natural connections to string theory. One perfectly good reason for working on string theory is to understand mathematics better. This is true both for the overt mathematicians and also the covert ones, who live in physics departments and further their interest in mathematics under the name of physics. Mathematicians do not enter mathematics because they want to understand quantum gravity. Perhaps they want to understand geometry better or to classify algebraic structures, but if their prime interest had been the quantum mechanics of gravity they would have studied physics rather than mathematics as undergraduates.

Deep issues of space, time and quantum mechanics do not get you a hearing in the geometry department. New ideas and methods for counting the number of distinct curves in Calabi-Yau spaces do, and these new methods arose from the study of how the equations of string theory depended on the geometry of the additional dimensions. As we saw in chapter 9, these techniques are now all packaged under the name of 'mirror symmetry'. These techniques worked and gave correct answers, but they were initially as mysterious to mathematicians as if they had been devised at a witches' sabbath. The problem of understanding how and why these techniques worked was one of the initial motivations for making many mathematicians interested in string theory. The ideas from string theory were foreign to mathematics, but better than the existing tools. Naturally, these mathematicians wanted to understand both how they worked and how they fitted into what was already known.

Mirror symmetry was simply the first avatar of what has become the sprawling field of physical mathematics. This area is in many ways part of mathematics. It involves the study of geometric structures and the relationships between them. It has the sense and feel of mathematics. It is not about describing the world as observed and is certainly not about predicting the results of laboratory experiments. However, in many ways it is also not traditional mathematics. The ideas that are taken as inspiration are drawn from physics, utilising the properties of quantum field theory (for example). The level of proof is also not at the level of conventional mathematical rigour; agreement in the values of two twelve-digit numbers, computed in entirely different ways, may be taken as convincing evidence for a result even though it is certainly not a proof in the mathematical sense.

We have also seen in chapter 10 the role of string theory as a factory of new ideas for looking for physics beyond that contained in the standard accounts of particle physics and cosmology. If string theory is true, it implies certain facts about the world. For example, the world contains extra dimensions. Even if not immediately observable, these extra dimensions will still leave some traces. One of these traces is the existence of moduli particles with extraordinarily weak interactions, whose couplings to familiar matter are only at gravitational strength. The behaviour of these particles in the early universe can lead to modifications of the background light of the universe, the cosmic microwave background. Another trace are types of particles called axions. The ideas of branes and strings and extra dimensions also provide many other possible scenarios that could be true and have observational consequences – although not discussed here, another example is the possible existence of gigantic strings that are not microscopic but instead stretch across all of the universe.

It would be foolish and false to claim that string theory leads to unique predictions for any experiments that are doable at the current level of technology. This does not prevent it from generating a plenitude of scenarios that are perfectly testable, in the conventional sense, at doable experiments. So far, all proposed extensions for physics beyond the Standard Model have failed. However for this topic, success does not depend only on the scientists involved. The laws of nature are what they are, and we do not get to choose when the next discovery will be. Good ideas may or may not be rewarded. The only approach is to continue looking, and to continue thinking of ideas that may show up in experiment.

It is for reasons such as these, and also many others, that string theory has grown so much. It is not that there are so many more people interested in quantum gravity: it is in many ways due to the applications to areas with no connection to quantum gravity that the crowds came and stayed. As we saw in chapter 11, string theory has had much success in quantum gravity, and there are still people who work on string theory only or solely because they want to understand quantum aspects of the gravitational force. However, to regard these people as a majority would be to paint an inaccurate picture of the subject.

It is in many ways the success of the applications to areas outside quantum gravity that have also caused string theory to be viewed as the standard approach to quantum gravity. It is not true of necessity that the correct theory of quantum gravity should provide deep and profound insights into subjects with little obvious connection to quantum gravity. In fact, there is no logical requirement for this at all. It could be that the correct theory of quantum gravity has nothing interesting to say about mathematics, is disconnected from the Standard Model, tells us nothing new about quantum field theory and offers no additional insights into theories of classical gravity. It is possible in principle that the theory of quantum gravity is a stand-alone entity that is disconnected from the rest of physics. This does not feel correct, but it is not logically excluded. However, to many it seems unlikely, and this feeling explains why string theory is so widely viewed as the best candidate idea here. The presence of so many insights beyond the sticker claim of quantum gravity makes it more likely, it is felt, that the sticker claim is correct.

14.2 RIVALS AND COMPETITORS

The same has not been true for other alternative theories of gravity. Examples are asymptotically safe gravity, causal dynamical triangulations or loop quantum gravity. These theories tend to have larger profiles outside physics than they do inside it. The outer profile is determined by scientists or journalists writing for a general audience; the inner profile is determined by technical results that many scientists care about. One of the reasons many people have heard about loop quantum gravity is that two of the scientists who invented it, Lee Smolin and Carlo Rovelli, have written books explaining their ideas for a popular audience. They are good authors. They write fluently, and by doing so ensure science is present in the public forum. However, it is neither unfair nor unjust to observe that they are not neutral commentators on the ideas for which they are the parents.

Why have these ideas not been so successful among scientists? I am going to focus on loop quantum gravity as one particular example. The positive arguments for loop quantum gravity primarily involve arguments based on various principles that any theory of quantum gravity 'must' have. For example, quantum gravity 'must' be background independent, quantum gravity 'must' respect Leibniz's principle of the identity of the indiscernible, quantum gravity 'must' respect the fact that spacetime has to be relational. On this view, string theory is unsatisfactory because it does not embody these principles. The fact that loop quantum gravity is said to incorporate these principles is claimed as a positive argument for this theory.

However most physicists – even most theoretical physicists – even most theoretical physicists working on 'fundamental' physics – are only mildly interested in quantum gravity, and they are even less interested in the philosophical principles it allegedly 'must' satisfy. If the discovery of quantum mechanics teaches one lesson, it is to be exceedingly humble about stating any principle nature 'must' satisfy. Most physicists are practical, and they like to calculate. Many have invested time learning string theory because it has helped them calculate the solution to a problem that were interested in solving, and they care far more about this than the correct answer to the question 'Is nature relational?'.

These other theories of quantum gravity also offer little outside quantum gravity. I had personal experience of this when a postdoctoral researcher at Cambridge. A leading advocate for loop quantum gravity came to give a seminar. He was talking about a proposal to realise the Standard Model of particle physics based on octopus diagrams. I am someone who has never been that excited by quantum gravity *per se* – the problems are too deep and too removed from experiment for my own taste. Each to their own, but I have always been deeply interested in the origins of the Standard Model and for ideas to go beyond it. The talk went on, and the speaker enthused, but both the talk and the questions revealed that the speaker was less than fluent with the intricacies of the Standard Model.

The Standard Model is part of the unquestioned core of particle physics, tested and known to be correct many times over. As said earlier, it is logically possible that someone could think their way to deep principles of quantum gravity even with zero knowledge of the quantum theories describing the other forces of nature. Personally, I do not expect this. This talk made clear to me both that loop quantum gravity was not saying anything interesting about the Standard Model, and also that the leaders of this area did not seem to know in full detail the Standard Model, its problems, and its open issues.

Learning a new area in physics takes time: probably at least a year or so to do well. Speaking for myself, why should I learn any given approach to quantum gravity? How much time to devote to it? I think that quantum gravity as a whole is a much deeper and harder problem than the areas of physics I know well. I am less willing to listen to someone on quantum gravity if their understanding of simpler matters, which I know are correct and I know I understand, is not as high as I would like it to be.

This is a personal response, specific to me. It may be just a function of my own experience and interactions. However, this is what substantially reduces my own motivation to invest significant time in understanding other theories of quantum gravity. The same is not true of string theory; for example one of the most passionate and formidable advocates of string theory, David Gross, has, as we saw in chapter 1, also won the Nobel Prize for foundational contributions to the Standard Model.

However – so what? Has not loop quantum gravity reproduced many of the required features of quantum gravity? Even if it lacks any connection to the Standard Model, there is no reason for every hard problem to be solved. If loop quantum gravity is able to produce some of the key features required of quantum gravity, this surely counts as compelling evidence that the theory is touched with truth and is on the right track.

One claimed example is the entropy of a black hole. As mentioned many times in this book, Stephen Hawking and Jacob Bekenstein showed in 1973 that a black hole has an entropy, given by one quarter of its area when measured in units of the fundamental Planck length. A key test for any theory of quantum gravity is to reproduce that factor of a quarter – after all, in quantum gravity this is a computable quantity, and by carefully doing a computation one should be able to derive this number by enumerating all the possible constituents that make up the black hole. It has been claimed that this calculation has been done in loop quantum gravity, and that the answer is indeed a quarter.

At first sight, this sounds extremely promising. However, the situation is actually less appealing than it sounds. On the one hand, reproducing the entropy of black holes is a superb test of quantum gravity – the theory *has* to reproduce that factor of a quarter, otherwise it is incorrect. On the other hand, this is a dreadful test of quantum gravity. The reason why this is a dreadful test can be understood psychologically. Suppose you have a theory that you really and deeply believe is correct, and you have to do a calculation to test the theory – *and you know in advance what the answer must be for your theory to pass the test, and it is just one simple number*. This creates all the wrong incentives. There are incentives to fudge the calculation. There are incentives to appeal to 'physical intuition' to paper over dubious steps in the calculation. There are incentives not to double-check and triple-check the calculation once you have the right answer. There are incentives to get the right answer because this will lead to a paper with lots of citations in an important journal.

It is important to realise that these incentives require no dishonesty at all, and I am certainly not suggesting any. It is just that tests where the person who knows the (simple) answer required, the person performing the test and the person most emotionally committed to the theory are all one and the same person are far from ideal.[1] The traditional check of science on this false incentive is experiment. The scientific gold standard has always been to make predictions in advance of the experiment, and then to see these predictions verified. Nature is gloriously indifferent to human prestige and human desire; it is what it is and its answers are what they are.

All research involving quantum gravity lacks this gold standard. The silver standard is to make predictions for theoretical problems, for which the answer is not known in advance but can be checked using very different techniques. This silver standard has been abundantly satisfied in string theory. This is through the applications, which we encountered in chapters 8 and 9, to both mathematics and the formal, exactly soluble parts of quantum field theory. The equation towards the end of section 8.2, on tests of the AdS/CFT correspondence, illustrates what is meant by this – no idle argument could ever reproduce that formula.

The silver standard has however never been attained for loop quantum gravity. The reason why loop quantum gravity – or indeed any other alternative theory of quantum gravity – has not had the success of string theory is that it has never solved someone else's problem. Loop quantum gravity has never solved an unknown problem, a problem that belongs to a different field and that can be formulated in an entirely different language. The only problems it claims to solve are those where the answer is either uncheckable or is, in a certain sense, known in advance. These claims are all controversial in themselves – for example, there is certainly no widespread agreement that

---

[1] For similar reasons, I believe that to a non-expert the most compelling tests of formulae for black hole entropy in string theory are not the original calculations by Strominger and Vafa, but rather the later ones that successfully match expressions for subleading corrections to the entropy. These subleading expressions have far more structure than the leading factor of simply one quarter, and so they are much harder to get right by accident.

loop quantum gravity leads to a calculation of black hole entropy – but they all involve known problems.

The reason for string theory's sociological success is the same as the reason for its scientific success: it has both formulated and solved unknown problems that were not even phrased when the subject was born. No one working on string theory in the 1970s could have foreseen the mathematical and other applications of the subject that would arise from the 1980s to the present day. In doing so, the string theory 'user community' has expanded to include many, many physicists who have absolutely no concern about quantum gravity. It is a theory that is of interest to those of wildly disparate interests.

None of this says that currently proposed alternative theories of quantum gravity should not have physicists working on them, or that any attempt should be made to prevent new proposals for quantum gravity. Much scientific success has come from funding smart people to work on what they want, how they want. For example, the United Kingdom's national science academy, the Royal Society, runs a highly successful scheme that funds promising young scientists to work on projects of their choice for up to eight years, giving them maximal freedom and minimal bureaucracy.[2]

This funding model is the venture capital approach to scientific research. Many paths are dead ends, but those that are not open up a thousand new routes. However – few scientists are interested in quantum gravity and fewer still will be convinced by your personal prior beliefs as to what quantum gravity *must* be like. Theories of quantum gravity have some similarities with religions – easier to found than to attract followers. It requires very good reasons to convince people to work on someone else's theory of quantum gravity.

A major theme of this book has been that string theory has succeeded because it has provided these reasons. It is many things to many people, and it has proved so much more than just an approach to quantum gravity. It is not simply for the true believers; it is a tool as well as an ideology.

14.3 PREDICTING THE FUTURE

What does the future hold for string theory? As the book has described, in 2015 'string theory' exists as a large number of separate, quasi-autonomous communities. These communities work on a variety of topics range from pure mathematics to phenomenological chasing of data, and they have different styles and use different approaches. They are in all parts of the world. The subject is done in Philadelphia and in Pyongyang, in Israel and in Iran, by those with every variety of opinion, appearance and background.[3] What they have in common is that they draw inspiration, ideas or techniques from parts of string theory.

---

[2]I declare an interest in this scheme; I am a fortunate holder of one of these fellowships, and they are fantastic.

[3]While personal enjoyment is not in the strict sense a good scientific reason, this adds to the fun of working in the subject.

It is clear that in the short term this situation will continue. Some of these communities will flourish and grow as they are re-invigorated by new results, either experimental or theoretical. Others will shrink as they exhaust the seam they set out to mine. It is beyond my intelligence to say which ideas will suffer which fate – an unexpected experimental result can drain old subjects and create a new community within weeks.

I can say with confidence that as mathematical results are eternal, the role of string theory in mathematics will never go away. It may wane or wax in fashion, but it will always be there. String theory is a consistent structure of something, and that consistent structure leads to interesting mathematics. These parts of mathematics are true in the same unqualified sense that the rest of mathematics is true, and they will always be true independent of what any experiment may ever say about the laws of physics.

The same is true about formal aspects of quantum field theory or gravity. Although they may be phrased in the language of physics, in style these problems are far closer to problems in mathematics. The questions are not empirical in nature and do not require experiment to answer. The validity of the AdS/CFT correspondence has been checked a thousand times – but these checks are calculational in nature and are not contingent on experiment.

What about this world? It is because of the surprising correctness and coherence of string theoretic ideas such as AdS/CFT that many people think string theory is also likely to be a true theory of nature. This comes from a prior belief that deep, interesting structures incorporating both gravity and quantum field theory are rare, and a rich example of such a structure is unlikely to be simply surplus to nature's requirements.

Will we ever actually know whether string theory is physically correct? Do the equations of string theory really hold for this universe at the smallest possible scales?

Everyone who has ever interacted with the subject hopes that string theory may one day move forward into the broad sunlit uplands of science, where conjecture and refutation are batted between theorist and experimentalist as if they were ping-pong balls. This may require advances in theory; it probably requires advances in technology; it certainly requires hard work and imagination.

However, the laws of nature are what they are, and it is not given to us to know whether the next major discovery lies either around the corner or a century hence. As I write this, the Large Hadron Collider has just recommenced operations with its second run, colliding protons at a new record energy of thirteen tera-electronvolts. As more data is collected, this run could lead to great discoveries – or simply more evidence for the Standard Model. We do not know. We can only look and see what is there.

There is no royal road to direct experimental evidence for string theory. The only way to proceed is by working hard, avoiding silly statements, and carefully exploring all possibilities – and through more data, always more data.

Notes and Bibliography

A reference of the form arXiv:yymm.nnnn or arXiv:hep-th/yymmnn refers to an article identification number on the online arXiv preprint server, currently located at http://arXiv.org.

Chapter 1

1. Gross and Wilczek's paper is *Ultraviolet Behavior of Nonabelian Gauge Theories*, in Physical Review Letters 30 (1973) pp1343–1346.

2. Politzer's paper is *Reliable Perturbative Results for Strong Interactions?*, in Physical Review Letters 30 (1973) pp1346–1349.

Chapter 2

1. An expenditure figure of £160 per person has been used. The number was taken from Scienceogram UK, http://scienceogram.org/in-depth/government-spending/, in September 2015.

Chapter 3

1. The text of the *Principia* is easily available online. The quotation is from the Scholium to the definitions, the wording from Motte's translation (1729).

2. Wordsworth's poem (1805) is *The French Revolution as It Appeared to Enthusiasts at Its Commencement* and is available online or in any edition of his works.

3. The quote from Schwinger can be found on p336 of *The Birth of Particle Physics*, ed. L. Brown and L. Hoddeson, Cambridge University Press (1983).

4. The history of quantum field theory (including the saying about infinity) is summarised in the first volume of *The Quantum Theory of Fields* (3 vols), Steven Weinberg, Cambridge University Press (2005).

5. Weinberg's paper is *A Model of Leptons*, in Physical Review Letters 19 (1967) pp1264–1266, and citation records can be found via the INSPIRE database, http://inspirehep.net.

6. The quote from Coleman can be found in the preface of *Aspects of Symmetry: Selected Erice Lectures*, Sidney Coleman, Cambridge University Press (1988).

7. The Matthean principle [Matthew 16:18–19] is taken from the King James Version of the Bible.

8. *'Ten Green Bottles'* and *'The Grand Old Duke of York'* are traditional British nursery rhymes.

9. Fermi is quoted in *More Random Walks in Science*, R. L. Weber, CRC Press (1982), although this may not be the original source.

10. Dyson's remark is reported in an interview with Frank Wilczek, *'Discovering the Mathematical Laws of Nature'*, in the *New York Times* of December 28th, 2009, currently available at http://www.nytimes.com/2009/12/29/science/29conv.html?'r=0.

Chapter 4

1. Goroff and Sagnotti's paper is *The Ultraviolet Behavior of Einstein Gravity*, in Nuclear Physics B266 (1986) pp709–736.

2. The Feynman quote is taken from *What Do You Care What Other People Think? Further Adventures of a Curious Character*, Richard Feynman, Penguin (2007).

Chapter 5

1. The quote from Joel Shapiro appears in his 2007 article *Reminiscence on the Birth of String Theory*, arXiv:0711.3448.

2. The quote from Lovelace appears in *Dual Amplitudes in Higher Dimensions: A Personal View*, at p199 of *The Birth of String Theory*, ed. Cappelli, Andrea et al, Cambridge University Press (2012).

3. Rutherford's quote can be found on p111 of *Rutherford and the Nature of the Atom* by E. Andrade, Doubleday (1964).

4. The quote from Susskind can be found in *Scientific American* 305, pp80–83 (July 2011).

5. The quote from Schwarz is in *Gravity, Unification, and the Superstring*, at p47 of *The Birth of String Theory*, ed. Cappelli, Andrea et al, Cambridge University Press (2012).

6. Peskin's quote is on p786 of the classic textbook *An Introduction to Quantum Field Theory* by Peskin and Daniel Schroeder, Westview Press (1995). I am taking the liberty of assigning the quote to the senior author of the book!

7. Green is quoted on p136 of *Superstrings: A Theory of Everything?* ed. P. Davies, Cambridge University Press (1988).

8. Weinberg's quote is on p223 of *Superstrings: A Theory of Everything?* ed. P. Davies, Cambridge University Press (1988).

9. The paper '*Vacuum Configurations for Superstrings*' by Candelas et al. is in Nuclear Physics B:258, pp.46–74 (1985).

10. The number 473 800 776 comes from the Kreuzer-Skarke classification in Adv.Theor.Math.Phys. 4 pp1209-1230 (2002), arXiv:hep-th/0002240.

11. Witten is quoted on p96 of *Superstrings: A Theory of Everything?* ed. P. Davies, Cambridge University Press (1988).

12. The paper by Bergshoeff et al. is Physics Letters B:189 pp75–78 (1987).

13. The paper by Duff et al. is Physics Letters B:191 pp70–74 (1987).

14. Duff's comment is from his 2015 paper '*M-history without the M*', arXiv:1501.04098.

Chapter 6

1. At the time of writing, all the talks for the Strings conferences for the past decade or so can be found online on the various conference web-pages.

Chapter 8

1. Maldacena's paper is Int.J.Theor.Phys. 38 pp1113–1133 (1999), arXiv:hep-th/9711200.

2. I thank Jeff Harvey for permission to include the words of the 'Maldacena'.

3. I have taken the expression for the anomalous dimension of the Konishi operator from the paper arXiv:1202.5733.

4. I particularly thank Andrei Starinets for advice on the historical parts of this chapter, and Dam Son for permission to include the text of the quoted email.

5. The quote from Kenneth Wilson is taken from a 2002 interview with the History of Recent Science and Technology Project, currently hosted at http://authors.library.caltech.edu/5456/1/hrst.mit.edu/hrs/ renormalization/Wilson/index.htm.

Chapter 9

1. Peter Goddard's comment can be found amidst his recollections in his 2008 article *From Dual Models to String Theory*, arXiv:0802.3249.

2. I thank Sheldon Glashow for permission to include his verse.

3. Greg Moore's essay *Physical Mathematics and the Future* is currently available on his website at http://www.physics.rutgers.edu/ gmoore/ PhysicalMathematicsAndFuture.pdf.

4. Hardy's 1940 essay *A Mathematician's Apology* is widely available and its full text can be found online.

5. Robbins' recollections appear in *New College: A History*, ed. J. Buxton and P. Williams (New College, 1979).

Chapter 10

1. A review of the cluster soft excess and the possible axionic explanation of it can be found in Angus et al., *Soft X-ray Excess in the Coma Cluster from a Cosmic Axion Background*, JCAP 1409 09:026 (2014), arXiv:1312.3947.

Chapter 11

1. The comment on the Thorne/Hawking bet comes from chapter 6 of Stephen Hawking's *A Brief History of Time* (Bantam, 1989).

2. The paper by Strominger and Vafa is *Microscopic Origin of the Bekenstein-Hawking Entropy*, Physics Letters B379 pp.99–104, (1996).

3. A useful reference on subleading logarithmic corrections to black hole entropy is *Logarithmic Corrections to Schwarzschild and Other Nonextremal Black Hole Entropy in Different Dimensions* by Ashoke Sen, JHEP04 156 (2013), arXiv:1205.0971.

Chapter 12

1. Ecclesiastes 1:9–10 is quoted in the King James Version of the Bible.

2. *Nobel Dreams: Power, Deceit and the Ultimate Experiment* by Gary Taubes (Microsoft Press, 1988) is an accessible popular account of the discovery of the W and Z bosons.

3. The OED records the first use of 'scientist' as by William Whewell in 1834: 'some ingenious gentleman proposed that, by analogy with artist, they might form scientist, and added that there could be no scruple in making free with this termination when we have such words as sciolist, economist, and atheist – but this was not generally palatable.'

Chapter 13

1. Feynman's opinion on philosophy can be found on p232 of *Surely You're Joking, Mr Feynman?*, Vintage (1992).

2. Max Born's words can be found in the text of his 1954 Nobel Prize lecture, available from the website of the Nobel foundation, www.nobelprize.org.

Further Reading and General References

Essentially all technical papers on the subject since 1992 are available on the electronic e-print, http://arxiv.org.

An accessible book for learning general physics is

The Theoretical Minimum: What You Need to Know to Start Doing Physics, Leonard Susskind and George Hrabovsky, Basic Books (2014).

Popular books on string theory are

The Elegant Universe: Superstrings, Hidden Dimensions, and the Quest for the Ultimate Theory, Brian Greene, Norton, 2nd edition 2010.

The Hidden Reality: Parallel Universes and the Deep Laws of the Cosmos, Brian Greene, Vintage, 2011.

The Little Book of String Theory, Steven Gubser, Princeton University Press, 2010.

Semi-technical accounts of string theory include

Superstrings: A Theory of Everything?, Paul Davies and Julian Brown (editors), Cambridge University Press (1992).

A Brief History of String Theory: From Dual Models to M-Theory, Dean Rickles, Springer (2014).

The Birth of String Theory, Andrea Cappelli, Elena Castellani, Filippo Colomo and Paolo di Vecchia, Cambridge University Press (2012).

String Theory and the Scientific Method, Richard Dawid, Cambridge University Press (2014).

The most accessible books for those who want to learn string theory at a technical level are

Basic Concepts of String Theory, Ralph Blumenhagen, Dieter Lüst and Stefan Theisen, Springer (2012).

A First Course in String Theory, Barton Zwiebach, Cambridge University Press (2009).

Lectures on String Theory, David Tong, arXiv:0908.0333.

Criticisms of string theory can be found in:

The Trouble with Physics: The Rise of String Theory, the Fall of a Science, and What Comes Next, Lee Smolin, Mariner Books (2007).

Not Even Wrong: The Failure of String Theory and the Search for Unity in Physical Law, Peter Woit, Basic Books (2007).

Peter Woit also maintains a well-curated blog of the same name, *Not Even Wrong*, at http://www.math.columbia.edu/ woit/wordpress/, covering a wide range of topics in physics and mathematics.

Farewell to Reality: How Modern Physics Has Betrayed the Search for Scientific Truth, Jim Baggott, Pegasus, 2014.

Books on alternatives to string theory include

Three Roads To Quantum Gravity, Lee Smolin, Basic Books (2002).

Covariant Loop Quantum Gravity: An Elementary Introduction to Quantum Gravity and Spinfoam Theory, Carlo Rovelli and Francesca Vidotto, Cambridge University Press (2014).

Index

Printed in the United States
by Baker & Taylor Publisher Services